HOW COMMUNITIES BUILD STRONGER SCHOOLS

STORIES, STRATEGIES, AND PROMISING PRACTICES FOR EDUCATING EVERY CHILD

Anne Wescott Dodd
and
Jean L. Konzal

HOW COMMUNITIES BUILD STRONGER SCHOOLS
Copyright © Anne Wescott Dodd and Jean L. Konzal, 2002.

Softcover reprint of the hardcover 1st edition 2002 978-0-312-23891-9

All rights reserved. No part of this book may be used or reproduced in any manner whatsoever without written permission except in the case of brief quotations embodied in critical articles or reviews.

First published 2002 by
PALGRAVE MACMILLAN™
175 Fifth Avenue, New York, N.Y. 10010 and
Houndmills, Basingstoke, Hampshire, England RG21 6XS.
Companies and representatives throughout the world.

PALGRAVE MACMILLAN is the global academic imprint of the Palgrave Macmillan division of St. Martin's Press, LLC and of Palgrave Macmillan Ltd. Macmillan® is a registered trademark in the United States, United Kingdom and other countries. Palgrave is a registered trademark in the European Union and other countries.

ISBN 978-1-349-63223-7   ISBN 978-0-230-60214-4 (eBook)
DOI 10.1007/978-0-230-60214-4

Library of Congress Cataloging-in-Publication Data
How communities build stronger schools: stories, strategies, and promising practices for educating every child / Anne Wescott Dodd, Jean L. Konzal.
    p. cm.
Includes bibliographical references and index.

    1. Community and school—United States. 2. Home and school—United States. 3. Education—Parent participation—United States. I. Konzal, Jean L., 1944– II. Title.

LC221. D6   2002
271.19—dc21

2002022040

A catalogue record for this book is available from the British Library.

Design by Letra Libre, Inc.

First edition: October 2002
10 9 8 7 6 5 4 3 2 1

Transferred to Digital Printing in 2013

To our husbands and best friends,

Jim Dodd
and
Bill Konzal

Contents

List of Figures and Sidebars vii
Acknowledgments xi
Preface xiii

Introduction xvii

PART I
Parent and Community Involvement Today: Challenges and Problems

1. Why Parents, Educators, and Community Members Must Work Together — 3
2. The Traditional Home–School–Community Relationship: Tensions and Limitations — 18
3. From Confrontation to Conversation: The Story of the South Wellington Parents' Special-Education Group — 42
4. Conflict about School Change: The Story of Springfield Regional High School — 70

PART II
Beyond Parent Involvement: Connecting Home, School, and Community

5. Changing the Paradigm: The Home–School–Community Relationship as Synergistic — 99
6. Getting Started: Building a Foundation of Trust and Respect — 138

PART III
Moving Closer to the New Paradigm: Profiles and Practices

7. A Place Where Everybody Knows Your Name: Greenbrook Elementary School — 181

8. Creating a Community of Difference: Crossroads Middle School — 203

9. Linking Home, School, and Community: A Sampling of Strategies and Practices — 231

10. Rethinking What It Takes To Educate All Children — 285

Appendix: Resources — 309
Notes — 320
About the Authors — 333
Index — 339

List Of Figures And Sidebars

FIGURES

Figure 2.1	Old Paradigm	25
Figure 5.1	Transitional Phase	109
Figure 5.2	New Paradigm	126
Figure 9.1	Parents' and Educators' Participation in the School as a Community for Learning	284

SIDEBARS

Sidebar 1.1	Why Educators Need to Listen to Parents—Vickie Gehm	8
Sidebar 2.1	Schools Make Unrealistic Demands on Some Families—Marilyn Brown	28
Sidebar 2.2	It All Comes Down to Communication—Marilyn Brown	32
Sidebar 2.3	Advocating for Children with Special Needs: A Parent's Perspective—Betsy Moyer	33
Sidebar 3.1	Teacher Recognition Award Bookmark	66
Sidebar 4.1	Springfield Regional High School Belief Statement	71
Sidebar 4.2	SRHS General Assessment Rubric	75
Sidebar 4.3	Grading and Reporting at Springfield Regional High School	76
Sidebar 4.4	Springfield Regional High School Governance Structure	92

Sidebar 5.1	Ten Trends: Educating Children for a Profoundly Different Future—Educational Research Services	101
Sidebar 5.2	A Letter to Parents—Scott Balicki	115
Sidebar 5.3	One Principal's Story—Margie L. Horton	122
Sidebar 5.4	The Principles of Community Schooling—Erwin Flaxman	129
Sidebar 6.1	Minority Parents Take the Initiative—Meredith Maran	147
Sidebar 6.2	Classroom "Socialism" or a Lesson in Sharing? from the *Bucks County Courier Times*	159
Sidebar 6.3	Reluctant Parents Participate in Arts Integration Days—Janice C. Marston	162
Sidebar 6.4	Parent Report Cards in Chicago—Julie Woestehoff	164
Sidebar 6.5	What Do Parents and Educators Want?—Kathy Teel Michel	166
Sidebar 6.6	Communication Is Key—Nancy Goldberg	169
Sidebar 6.7	Claire's School—Jean Konzal	172
Sidebar 6.8	The Essential Advisor—Larry DeBlois	175
Sidebar 7.1	From the Principal's Desk: *Greenbrook Community News*	183
Sidebar 7.2	A Typical Year at Greenbrook, 2000–2001 Calendar	187
Sidebar 7.3	Mystery Readers—Lori Woods	195
Sidebar 7.4	Home Visits—Lori Woods	198
Sidebar 7.5	Parent Conferences—Thea Dahlberg	200
Sidebar 8.1	A Parent's View: Black History Month Must Be Celebrated in Our Schools—Carolyn Jordan	204
Sidebar 8.2	The Director's View: Multicultural Curriculum: What Is It? How Do We Do It?—Ann Wiener	205
Sidebar 8.3	Crossroads School Calendar, 2001–2002	221
Sidebar 8.4	Crossroads Mission Statement	223
Sidebar 8.5	Crossroads School Parent Statement—Marjorie Moore	226
Sidebar 9.1	The More, the Merrier: A Successful and Effective Strategy for Building Bridges of Communication—Kelly Arsenault and Debbie Deschambeault	233

Sidebar 9.2	Connecting with My Parents—Susan F. Bean	236
Sidebar 9.3	E-Mail: A Powerful Way To Communicate—Julie Ann Ludden	238
Sidebar 9.4	Mini-Schedule—Larry DeBlois	239
Sidebar 9.5	Bridges—Kathy Gregg, Brendan Dundas, Nancy Fowler, and Sue Manning	244
Sidebar 9.6	"Mom and Pop Reading Day"—Pat Monohan	254
	"Mom and Pop Reading Day" from a Parent's Perspective—Julie Macosko Kerber	
Sidebar 9.7	Parent Involvement in the High School—Chris Horwath	256
Sidebar 9.8	Parent Volunteers Create a Win-Win-Win Situation!—Nancy Goldberg	257
Sidebar 9.9	Parent Workshop—Deanna Nadeau	258
Sidebar 9.10	Sharing Writing with High School Parents—Holly Ciotti	260
Sidebar 9.11	"Math with Mom"—Deanna Nadeau	262
Sidebar 9.12	"Discovery with Dad"—Deanna Nadeau	263
Sidebar 9.13	Bringing Parents and Children Together through Poetry—Elaine M. Magliaro	264
Sidebar 9.14	Sharing Teaching Philosophy with Parents—Karen M. Christian	266
Sidebar 9.15	Parents in Partnership: Freeport Middle School—Janet Russell Theriault	267
Sidebar 9.16	Engineers Help Students Build Robots—Kit Juniewicz	274

Acknowledgments

As is true of other books, we could not have written this book without the assistance of many other people, including many individuals whom we do not name below. We thank them all.

Although we have used pseudonyms for South Wellington and Springfield in the book, these are real schools and communities with wonderful people who were generous enough to share their time and personal perspectives with Anne. Thus, we thank the parents and administrators in both communities for their valuable contributions, which helped us understand how parents and educators can move beyond conflict when the adults find ways to work together for the benefit of children.

We are appreciative of the help provided by the students, faculty, staff, and parents—past and present—of Crossroads School in New York City who warmly welcomed Jean into their midst. Special thanks go to the school director, Ann F. Wiener; teacher Rebecca Hendrickson; office manager and parent Ana Chanlatte; and parents Harriet Bograd, Carolyn Jordan, Marjorie Moore, J. E. Miles Jr., and Leslie Flack Newman, who, in addition to participating in interviews, read and commented on the many drafts of the chapter.

Special recognition must go to the children, parents, teachers, and principal of Greenbrook Elementary School. For five years coauthor Jean Konzal has worked closely with them to supervise student teachers and to facilitate the Kindergarten Parent Journal Writing Project. She thanks the teachers for making her an honorary member of their faculty—including her in their many discussions about teaching and learning and about the day-to-day problems that arise when working with children. In addition, she thanks all the parents involved the journal writing project who helped her better understand their concerns, but especially Shari Rothstein, Carol Desmond, and Miriam Lally, who wrote short essays about their relationships with the school that are quoted in the book; teachers Thea Dahlberg, Lori Woods, and Ellen Gordon, who contributed material for this book; and especially the principal, Pat Holliday, who makes all that happens in Greenbrook possible by creating an open and nurturing climate where children, their parents, and teachers thrive.

We are grateful to our institutions—Bates College for various kinds of support for Anne and The College of New Jersey for providing released time to Jean during the Fall 2001 semester through the Support of Scholarly Activities fund, and to several students: Jesse Reich, a 2001 Bates graduate, for research at Springfield Regional High School; and student assistants David Weliver, Katie Burke, J. J. McGill, and Kristen O'Toole, students at Bates College, and Jennifer Stevens from The College of New Jersey.

In addition to all of the contributors of sidebars to the book, we thank Kit Juniewicz, who not only wrote sidebars herself but also went the extra mile to convince others to contribute.

We appreciate the work of our editor, Michael Flamini; his assistant, Amanda Johnson; and the many others at Palgrave Macmillan, who helped to get the manuscript ready for publication. We are especially grateful to Michael for his faith in the quality and value of our work—as well as for introducing us to some great restaurants we would never have discovered on our own!

Finally, we could never have done this project at all without the support of our families. Anne thanks her daughters, Vickie and Suzan, and her six grandchildren, Kristen, Cori, and Casey in Florida, and Eric, Marcus, and Sara in Massachusetts, for helping her understand schools from multiple perspectives. She has learned a great deal from listening to their experiences (some positive, some not) with schools in different states over many years. Anne thanks her husband, Jim, for giving her "private time" to work on weekends by getting out of the house to go sailing—even in the winter. Jim has also been incredibly supportive of her professional work in other ways, including reading drafts to point out where the jargon was too heavy. Most of all, she appreciates his unconditional love and friendship.

Jean thanks her husband, Bill—a.k.a. Wild Bill—who has been a constant source of encouragement and perspective. As a superintendent of schools, his real-life view of schools kept her grounded in reality. His sense of humor kept her laughing and, most important, his unqualified belief in her ability sustained her in times of doubt. Her son and daughter-in-law, Gregory and Angela, and their three children, Eliot, Noah, and Michael, have provided her with the reason to persevere. As she watches her three grandchildren grow and learn, she is even more committed to making schools exciting places for all children, places where, as Eleanor Duckworth says, children are encouraged to "have wonderful ideas."

Preface

Our children need us. . . . Adults themselves need to come together. The same child that is being taught in the school is being served by the judge, by the municipal health people, by the police, by all the community-based organizations. We've got to form a common ground so we can unify what we are doing. Our kids need us. We need our kids so we must serve them better.

—U.S. Commissioner of Education Rod Paige
in a satellite town meeting discussion broadcast
shortly after the tragic events of September 11, 2001

Most of this book was written well before the four planes crashed into the Twin Towers of the World Trade Center in New York, the Pentagon, and a Pennsylvania field on September 11, 2001. These unimaginable events have forced all of us to look at our country, our lives, and our children in very different ways. Parents and teachers have had to explain what happened to children, who want to know why anyone would want to kill so many innocent people, when they have no understanding themselves. In fact, some of us may still find it difficult to believe that what happened on that awful day was not just a bad movie.

We carry with us, no matter where we go or what we do, images of the destruction at Ground Zero—and fear and uncertainty that won't go away: about ourselves, our loved ones, and our futures. We want to help so we give blood in record numbers and donate to organizations to help the families of the victims and to funds established for scholarships in memory of those who died. Even without knowing what can be done to prevent future attacks, we want our country's leaders to do something.

Although it is always difficult to see that anything good can come from something so bad, we have seen people in all areas of this country come together for the first time in many decades. People sing "God Bless America," wear red, white, and blue ribbons, and fly American flags, small and large. No matter what our age, race, ethnicity, political beliefs, or personal values, we have something in common. This shared experience has given us the means to begin a conversation with anyone, anywhere. And if this feeling of unity doesn't dissipate too soon, it can be a catalyst for us to work together to solve other problems no less urgent to the preservation of our democratic way of life than the challenge of eliminating terrorism.

If we want our children to live in a safer world tomorrow, we must look at how we are educating our children today. We need to recognize that *education* doesn't just happen in schools but is going on all the time in all aspects of our children's lives. And teaching children is much more complicated than telling them what they need to know and what they should or should not do: The most powerful and long-lasting lessons children learn come from what they see us do and say. The golden rule, "Do unto others as you would have them do unto you," means little if children hear their parents say "Those people (Mexicans, Muslims, Chinese, or others who are not white) should go back to their own country," "Stay away from those kids (whose skin is a different hue or whose clothing is dirty and torn)," or "If those Arabs are going to live in this country, they should try to fit in, wear clothes like us, and learn to speak English." Or if in the grocery store children watch an adult move quickly away from a woman whose head is covered with a veil or see someone stare at a disabled young boy, trapped in a wheelchair and unable to keep from lolling his head. Our actions, as the old saying goes, speak louder than our words.

One of the greatest benefits of a strong public system of education in a country so diverse as ours—and growing more diverse all the time—is that children have the opportunity to learn with and from children whose backgrounds are very different from their own. Yet advocates of school reform argue that public schools today are not teaching all children well. Politicians and scholars have many different views about what is wrong and what we need to do to "fix" schools. As the authors of this book, Anne and Jean do not take a position for or against any particular "solution" since there probably isn't one "right way," but many possibilities. What we do know is that people in each community must decide for themselves what will help their own children learn.

When we began this book, we thought we would research and share some ways parents and educators could work together to improve public schools. As we got deeper and deeper into the complex issues facing public schools that serve children with so many different needs, we realized that, just as no parent or teacher alone could help a child understand what happened on September 11, 2001, the likelihood of transforming schools into places where *all* children learn would be greater if everyone—parents, teachers, community leaders, and everyday citizens—worked together, as U.S. Commissioner of Education Rod Paige says, "to unify what we are doing" because "our kids need us. . . . so we must serve them better." Thus, in this book we offer a new way of thinking about the relationship among home, school, and community, one that integrates these three important aspects of every child's education. We contrast this new synergistic model with a traditional view of home, school, and community as separate satellites, operating independently or with only tenuous or problematic connections to each other.

The events of September 11 may suggest that the synergistic paradigm should be extended on a global basis. The more we know about the perceptions and beliefs of others, the greater the chance we will have for working together for a common purpose. That is probably the main point of this book: When finding common ground eludes us, we can work in different ways to achieve a common purpose.

In this book we explain why and how we should rethink what it takes to educate all children well, and we share some stories and suggest strategies that point the way to an admittedly idealistic goal. And if there are some readers who do not think this work is important, they should consider that every drug addict, every mass murderer, and every terrorist was once a small child and later a student in a school and ask themselves: How might our country (and the world) be made a safer, more peaceful place?

Introduction

Our first book, *Making Our High Schools Better: How Parents and Teachers Can Work Together* (St. Martin's Press, 1999) focused only on high schools and only on the relationship between parents and educators. When we began this book, we intended to build on our previous work by researching and sharing strategies for involving parents with schools from kindergarten through grade 12.

At some point in this process (we aren't sure exactly when this happened or why), however, we both began to realize that we were moving beyond traditional notions of parent involvement to new ways of thinking about the parent–school relationship. We also could see that improving public schools depends not only on parents and educators working together but must include everyone else in the community—people who do not have children in school but support schools through their taxes, business owners who employ students when they complete their schooling, and community members whose quality of life is enhanced, as when graduates become productive citizens and workers, or diminished, as when school dropouts are unemployable or engage in criminal activities. Thus, to educate children effectively, schools need to build strong connections with parents *and* community members.

As we continued to collect stories about schools and strategies for involving parents and community members with public schools, we soon recognized that the challenge of educating all children well in the United States today is so complex, so nested in all aspects of society, that the public schools need a great deal of help. In fact, because children are learning all the time in every context of their lives, the community—not just the school or their parents—is responsible for their education. Yet because all children have access to public schools and because everyone must support them, we began to rethink the existing relationships among home, school, and community. Could we improve the education of all children if we imagined what might be possible if the connections among home, school, and community were seamless and synergistic? Many people in many different ways are already serving the needs of children, but their efforts, often fragmented and disconnected, duplicate—and sometimes even contradict—what others are doing for the same children.

These important insights, which may seem like common sense to some readers, are not new. We are indebted to Joyce Epstein, James Comer, and many others whose research is more extensive and well known than ours. Their work is the foundation for what we call "the new paradigm," a seamless connection among home, school, and community. The new paradigm proposed in this book is an ideal. We ask readers to suspend their disbelief that such an ideal is ever possible to consider the potential benefits of using this way of thinking about these relationships as a catalyst for discussion and dialogue and as a guide for action. We believe that having such an ideal in mind—even one that may never be possible in reality—is an important first step in serving *all*—not just some—of our children well. A democratic society should do no less.

As readers will see, we focus more on the relationships between parents and educators—even as we argue that including community members in the education of children is essential. Some readers may think that we contradict our own purposes by paying less attention to the community than the new paradigm suggests would be ideal. There are several reasons that schools should emphasize the strengthening the home–school relationship. First, because of lack of time and resources, educators can't do everything at once. While they shouldn't ignore the wider community, they need to set some priorities. Second, when parents, who are also community members, work more closely with educators, they can become the links, or liaisons, to other community members and resources. Finally, because parents have the strongest personal investment in what schools do for—or to—their own children, they have the greatest motivation to be more involved. Educators, then, can "practice" skills and strategies for building trust and respect with parents before they later adapt them to work with others in the community who have no children in schools. Moving toward any ideal is both a developmental and an incremental process. So we suggest that beginning with a focus on parents is a sensible first step. However, we do not want to suggest a linear, step-by-step process. The process may be more like a spider's web than a straight line. Even as educators build relationships with parents, they can simultaneously find opportunities to reach out to the larger community.

Part I of the book focuses on the challenges and problems of parent and community involvement in schools today.

Chapter 1 considers philosophical, moral, and practical reasons for establishing closer connections among home, school, and community. Then, in chapter 2, we present what we call "the Traditional Paradigm"—home, school, and community as separate, independent satellites—and the tensions and limitations it creates. The school in this model is like a medieval fortress: Educators try to keep parents and community members from interfering with what they, as the professionals with expert knowledge, are doing to teach children.

Because most existing home–school–community relationships have many of the characteristics we attribute to the traditional paradigm, conflict with parents and community members is very likely. Yet as the stories told in chapters 3 and 4 show, conflicts, although painful for everyone involved, can result in some positive outcomes. Special-education parents in South Wellington (a pseudonym) who confronted school officials about the failure of the schools to serve their children ended up working with them to improve education for all children with special needs.

In Springfield (also a pseudonym), educators struggled to get a brand-new school with a totally different standards-based curriculum and grading system and untracked classes up and running. Because these practices were so unlike those in traditional high schools, parents and community members organized a protest that led to voters in June of the school's first year being asked to signal their support or opposition to the new practices and to the top administrators (whose names all appeared on the ballot). As chapter 4 shows, however, despite the tumultuous first year, Springfield Regional High School experienced a quiet second year without making major changes—except in its relationships with parents and community members.

In Part II, we present a new way of thinking about these important relationships.

As chapter 5 explains, changing the old paradigm is a developmental process that requires commitment and creative thinking. First, we show that scholars and others in many disciplines are thinking in ways that parallel the new paradigm of home, school, and community relationships we propose as an ideal. Then we describe what we call "the transitional phase" with examples to show that the changes we envision already exist in many places. The chapter ends with an explanation of the synergistic model and questions as catalysts to simulating readers' imagination about how things might be different—and how such changes would benefit our children.

Because all positive relationships require a foundation of trust and respect, chapter 6 discusses this topic in some depth. It considers the different expectations parents and educators have of each other and the barriers that prevent them from trusting each other, and suggests some ways to begin the process.

Part III, except for the conclusion, focuses on practical ways to move closer to the ideal.

Chapters 7 and 8 profile two schools, Greenbrook Elementary School, a suburban school in New Jersey, and Crossroads School, a diverse, urban middle school in New York, that have made significant gains in involving parents with the school. These schools also show that strengthening parent–educator relationships makes possible the opening of doors to the community. If educators and others are looking for additional practical ideas beyond those suggested in the profiles, they will find many more in chapter 9, which provides a sampling of practices and strategies from schools and communities across the nation.

Finally in chapter 10 we consider the implications of our work for rethinking what it really takes to educate all children well.

The appendix lists a variety of resources—organizations and websites—that provide additional information for readers who want to move from thought to action. There is no more important challenge for all of us than finding ways to serve all of our children well. Nothing that we do has a greater impact on the future than what we do for children today.

Part I

PARENT AND COMMUNITY INVOLVEMENT TODAY: CHALLENGES AND PROBLEMS

1

WHY PARENTS, EDUCATORS, AND COMMUNITY MEMBERS MUST WORK TOGETHER

Children are living messages we send to a world we will not see.

—Amish proverb

Model Ts and outhouses are relics of an American past, but that time seems, to some people, to have been a golden time. Kids listened to their parents and did what teachers asked them to do. Mom, Dad, and the children ate together at home every night. Schools focused on teaching the three Rs: Every student memorized the times tables and knew how to spell. Clearly, the country was better off when the "F" word was never uttered in public, a McDonald's or Burger King couldn't be found in nearly every town, and phonics—not "whole language"—was the preferred way of teaching students to read. Even if this "golden time" ever did exist, we can't go back. Our children now live in a world of computers, video games, MTV, nonnuclear families, global competition, and, since September 11, 2001, terrorist attacks in our own country. Changes, which used to plod into small towns on horseback, seem to arrive now on sleek jets and, thanks to technology, quickly impact communities everywhere. Schools today are faced with the challenge of preparing children to live in a future we cannot predict with any certainty but one: Change will be a constant.

Someone once said that the only one who welcomes change is a baby with a wet diaper. Change is problematic because it means something new, and most of us fear what we do not know. Yet this aversion to change is paradoxical because one of our most cherished democratic ideals is based on change: Every individual in this country should have the freedom and opportunity to better him- or herself. We love Horatio Alger stories. We admire those who work hard and achieve a great deal: the inventor who tests new ideas in a garage and develops a product that transforms American life, the immigrant who arrives speaking no English and later becomes a college professor, the poor children who pay their own ways to college at night and later become prominent and powerful. Such accomplishments require change. The problem is that no one can be certain what changes are needed or when a given change is an improvement. Some changes have made things worse, not better, often because no one considered the unintended consequences—until they became new problems to solve.

When any change is introduced anywhere, it is likely to generate conflict. As we have all seen many times, the more radical the change, the more people affected by it, the more widespread, disruptive, and painful the reaction against the change will be. Because public education affects people everywhere, discussions and decisions about change in schools create a public arena for conflict—or cooperation. School administrators may wish that parents and community members with different ideas, different perspectives, would go away and let them do their jobs, but the reality is that, in a democracy, public schools are the public's schools, and what schools do—or do not do—makes a difference for everyone. The education we give our children today is the foundation on which we are building the kind of world we want to live in tomorrow.

Although some educators would like to control or limit parent participation in schools, everyone agrees that parents should be involved with their children's education. In fact, parents and others in the community can't be "not involved" if one considers that *education*—in contrast to the more limited concept of *schooling*—encompasses everything children learn simply by living in this society. Parents are, after all, their children's first and most important teachers. In a democratic society, however, because public schools serve the common good by providing the knowledge, skills, and experiences children will need to function effectively as adults in a complex and rapidly changing world, most people also would agree that *schooling* is essential as a part of children's broader educational development.

Parents and educators share the same important goal: to help children become successful in school and in life. This goal is important to the wider community as well: Business owners need competent employees, communities need informed citizens who will consider the public good as well as their private interests when they cast their ballots, and no one wants to see taxes increase to pay for building more prison space. But even though almost everyone agrees with this lofty goal, there is no common agreement about the process for attaining it.

Since the publication of *A Nation at Risk* in 1983, the report on the state of American schools that warned everyone about "a rising tide of mediocrity that threatens our very future as a Nation and a people," there have been many, various efforts undertaken to improve public schools.[1] Almost two decades later, the debates about what is needed to improve schools continues. Should we set standards for learning that every student must achieve before graduating from high school? Should we make all schools (and teachers) more accountable through greater use of standardized tests? Should we increase the requirements for teacher preparation and certification? Should we give parents vouchers so that they can choose to educate their children in private schools? Should we encourage the establishment of more charter schools that have greater freedom to experiment with innovative practices? There is no shortage of ideas from all sides of the political spectrum, but agreement continues to be elusive.

Regardless of what any community chooses to do to improve local schools, educators who hope to make changes that will last must involve parents in the process. "The role of the parents in educational reform," says noted educational researcher Michael Fullan, "has been both sadly neglected and underestimated.... [S]trategies to involve parents represent one of the most powerful and underutilized instruments for educational reform."[2] Other citizens also must be involved in these efforts because, as any experienced school administrator knows, any change is likely to generate some conflict. Too great a change without sufficient parent and community involvement can become a full-scale crisis. The losers are always the children who get caught in the middle of warring adults.

The future of the American society depends on the quality of its public schools. American children deserve no less than the best education possible. While there will never be 100 percent agreement on what "best" means in practice, everyone must work together. Schools can—and should—become better at helping all children grow into successful adults and productive citizens. What "living message" do we as a society want to send to the world we will not see?

WHY PARENTS AND COMMUNITY MEMBERS SHOULD BE INVOLVED WITH PUBLIC SCHOOLS

What the best and wisest parent wants for his own child, that must be what the community wants for all of its children. Any other ideal is narrow and unlovely; acted upon, it destroys our democracy.

—John Dewey, *The School and Society*

To accomplish these societal goals, most people would agree that parents and the public should be involved with schools for both idealistic and pragmatic reasons. Yet parents' rights and responsibilities, when considered in the context of real issues and situations, raise important questions no one can answer with any certainty: What is a "good" school? What school practices are "best"? Who should decide?

The Democratic Ideal

The democratic ideal requires that all citizens have the opportunity to have their voices heard and to share in the public decision-making process. Although teachers may prefer that parents not have much say about the school's curriculum, for instance, educators in public schools must recognize that in a democratic society schools—and, by extension, their practices and programs—exist to serve—not to control—parents and citizens. However, because people do not share the same values and beliefs, discussions about educational policy and changing school programs and practices will also be a forum for the expression of competing points of view. Even though such discussions may become emotional, parents need to be involved in the process especially when a new program or practice will impact their children directly.

The democratic ideal recognizes that parents have an important role to play. In fact, many people would argue that parents have a moral obligation to support and encourage the personal, social, and academic development of their children. They also have both a right and an obligation to make sure that children are well served by the schools they attend. Parents are both teachers of their children and mediators of the school culture.[3] As models for learning, they exert a powerful influence on children's ability to learn in school.[4] Researcher Robert Friedman notes, too, that "negative feelings about the family's previous experience with a teacher, a school, or the 'educational establishment' are picked up by the child and acted on.... Home standards that are contradictory or inconsistent may result in the child's nondeliberate distortion of class rules."[5] Another researcher points out that "behavior in a classroom is often a carryover of home experience."[6]

If these researchers are right—and other studies would suggest that they are—then educators need important information and assistance from parents to resolve school problems. Connecting and communicating with parents about school matters is not just a matter of paying lip service to parents' right to be involved; doing so will enable teachers to do their own work more effectively.

Parental Involvement and School Achievement

Student achievement is another reason parents should be involved in schools: Research shows that parent involvement influences the level of student achievement. For example, a U.S. Department of Education publication citing eight research studies for support states that "Studies of individual families show that what the family does is more important to student success than family income or education. This is true whether the family is rich or poor, whether the parents finished high school or not, or whether the child is in preschool or in the upper grades."[7] Parent involvement can also build support for the school and its staff because parents who are involved in school are more likely to have more positive opinions about teachers.[8] The findings of this study by Carol Ames and others also suggest that the frequency and content of school-to-home communications are important. When these communications contain information that influence parents' perceptions of their child as a learner, when they give parents a sense of efficacy, and when they make the parent feel comfortable with the school, they enhance parent involvement.

Parents and School Reform

Parents need to be involved in change efforts because their perspectives affect the success of any reforms a school attempts to implement. When parents or educators have opposing points of view about a particular issue or

---

Sidebar 1.1

Vickie Gehm
Parent, volunteer, and substitute teacher
Melbourne, Florida

Why Educators Need to Listen to Parents

Lori was in the fourth grade before her father and I could convince any of her teachers that she had a learning problem. We finally got her teacher to recommend that she be tested for a learning disability, but the principal refused because her grades were too high. Because she had all Bs and above, the principal thought there couldn't possibly be a problem. We were told that Lori was a very bright student, hardworking, a good example for her classmates, and nice to have in the class. However, the story at home was much different. She hated reading. Although she liked working with numbers, math (her favorite subject) was becoming very difficult. Homework was becoming more and more frustrating for her. When we tried to help her at home, it was a nightmare. We all went to bed frustrated, exhausted, and angry.

As a testament to her intelligence, Lori had figured out ways to hide her problem. For example, she would work with the smartest kids in class on homework assignments before school got out. (This meant she didn't have to work with us at night!) Basically she would let them do the work, and they would give her the answers. After the fourth grade I don't think she ever read a book. She was required to read several but read only enough to complete the book reports or read books her friends read and asked them to tell her about the book. Spelling tests weren't a problem because Lori had a good memory (although the words were quickly forgotten once the test was over) so her test grades were very good. If she misspelled words in her daily work, the teachers didn't see that as a problem: She was only in the fourth grade, and, after all, we can't all be good spellers.

Finally in the fifth grade Lori got a teacher who saw the problems she was having staying on top of things. Since Lori was older, more was required of her, and she was having trouble hiding her problems and blending in. She was required to do more individual work, more research, harder math. As the year progressed, her teacher saw her grades and her class participation drop. Lori's confidence and self-esteem were clearly shaken. When her teacher approached us about Lori, she seemed surprised that nothing had been done to help her. She wondered why she hadn't been tested. Did we not want her tested? We told the teacher we had wanted her tested since second grade, but the only test done was an IQ test. Since she scored high on that test and her grades were good, no one felt other tests were needed.

Our plan was to have her tested now. Once again, the principal at first disagreed, but we and her teacher insisted. We all were surprised when we learned Lori had done so well on the test she was given that she didn't fit the criteria for a learning disabled (LD) student. How was that possible? The principal herself

(continues)

> **Sidebar 1.1** *(continued)*
>
> tested Lori and felt that Lori had no learning disability. Further questioning revealed, however, that she had tested Lori orally.
>
> This result suggested to us that Lori had a processing problem. She knew the answers but couldn't get them from her head to the paper correctly. Although the principal refused to designate Lori as learning disabled, Lori's teacher saw that she needed more help. Lori would get so frustrated—she spelled words wrong on paper, wrote math problems backward, and switched numbers all around—that she would just give up. Lori's fifth-grade teacher decided that, if oral tests were good enough for the school to use to test Lori, then they were good enough for her to use with Lori, too. She would allow Lori to retake immediately any written test she failed as an oral test. The teacher was patient, understanding, and encouraging. We all saw immediate improvement. Lori's grades improved, and she was returning to her old self-confident, curious, fun-loving self. She enjoyed learning again.
>
> Unfortunately for Lori, though, the sixth-grade teacher didn't agree with methods used by the fifth-grade teacher. Thus, once again, Lori struggled. Then, when Lori entered junior high (with a different principal in charge), she was tested again. This time she was easily diagnosed with a "processing disorder." Now Lori is in eleventh grade, taking LD math and science classes along with regular high-school English and economics. She also has been given the opportunity to dual enroll in the local community college in a cosmetology program. Thus, she goes to high school each day from 8 to 1 and then attends classes at the college from 3 until 7:30.
>
> Even though Lori now has to work harder and longer than any of her friends, she is doing very well. She is learning, and she is happy. As parents, though, we can't help but wonder how things might have been different a long time ago if all those people in the school had listened when we first tried to tell them something was very wrong.

curriculum. This study revealed that parents blame the innovation for children's problems even when the real reasons for students' lack of success are unrelated to the new program or practice.[9]

When educators consider changes in the ways school has traditionally been, they must not forget that parents do not want their children to be used as guinea pigs in educational experiments. Because they will need parent and community support for the successful implementation of a new program or practice, they need to demonstrate how a planned change will benefit students—and help parents come to believe that this innovation will help rather than hurt their children. When parents perceive that a change is not an improvement, they will block it or create conflict that saps the energy and commitment of educators to continue.

Parental and Community Political and Financial Support

As a very practical matter, parents and citizens' opinions about schools also make a difference because they affect the level of financial and political support in a given community. As taxpayers and voters, parents and the increasing number of people who do not have children in the school are the ones who decide whether to increase or decrease school funding. In some communities voting to approve the annual school budget is direct—up or down, yes or no. In others, voters elect representatives who make those decisions, and, as taxes continue to rise, many enthusiastically support the candidates who promise to keep costs and property taxes down. If people do not like what the schools are doing, they not only say no to new spending but sometimes organize with others to cut the school budget. In one small town, where decisions were made at often-emotional town meetings, the entire school budget was reduced by one-third because people were unhappy with the changes in the philosophy and practices introduced during the past year at the local high school.

Voters' opinions about school practices also matter when there are bond issues for updating old school buildings or for new construction. Researcher Jack Wilhite found that lack of community support for school buildings that would facilitate the adoption of innovative practices was a problem. Educators in the rural community he studied said that it was difficult to implement innovative practices, such as team teaching, in existing facilities because the buildings had been designed for traditional teaching in order to win the support of voters. In this regard, Wilhite also points out that "[c]ertain voices in the community are heard by school officials; and, consequently, in a very direct way, they influence what is sometimes taught and the way it must often be taught."[10]

Public opinion and the level of general and financial support for schools in a community impact everyone in less obvious but no less important ways. Educators should routinely—not just when there is an upcoming bond issue or potential problem—give ongoing public communication efforts a higher priority. They need to go beyond providing information of interest to parents and getting the local press to cover school events by thoughtfully planning how they can communicate what the school is doing to the community. Local property values, for example, depend on the reputation of schools in the wider world. Houses in places where schools and teachers are held in high regard sell for higher prices than those in similar communities where public schools are seen as mediocre or deficient.

People form opinions about schools whether they have any direct contact with them or not, perhaps from informal conversations with neighbors who have children in school or from what they overhear people saying in the grocery store or barber shop. Perception is reality, and school stories—positive or negative—pass from one person on to another. Like the old school game where one person whispers a sentence that gets passed around the class and ends up totally changed, these school stories may bear little resemblance to reality. Because every direct contact a principal or teacher has with a parent will be multiplied as many times as that person shares his or her experience with others, educators need to make sure their interactions with all parents end on a positive or at least a neutral note rather than a negative one. One study, for instance, revealed that parents who had negative experiences were more likely to report having talked to other parents than did parents who saw the schools in a more positive light.[11] The fewer parents who leave a meeting disgruntled, the fewer people in the community there will be to hear what's wrong rather than right with schools. This doesn't mean, however, that educators should force their views on parents and community members. Instead, educators need to promote two-way communication and provide venues for really listening to the concerns people have. They must take them seriously and be honest with parents and the public about the challenges and problems.

Because educators do not have all the answers, they need to think not only about "public involvement" but also about "public engagement." They should create opportunities for parents and others in the community to join with them in clarifying the mission for public schools in a given community: What do our children need to know and be able to do when they graduate from high school? The discussions should not stop with the development of a "vision statement" for schools, as has been done in many places. People generally will agree with an outcome when stated in general terms, but the devil is in the details. The public, especially parents, also needs to understand the practices a school will adopt to meet the goal. Take the statement "Students will be effective communicators," for instance. It is nearly impossible to imagine that anyone would disagree with this outcome. But, when one thinks about how teachers will work with students to reach this goal, the unanimity can disappear in a hurry. Some parents, for instance, may expect teachers to give spelling tests every week and mark every misspelled word on all written work that students do. Teachers, though, may believe that spelling is best taught in context and that red-circling errors on students' early drafts will hinder them from learning how to express their ideas. They may want students sometimes to use writing as thinking on paper when the point is not correct writing but clearer thinking. Thus, they may decide to mark spelling errors only on some papers, final copies that students produce after multiple drafts, revised, corrected, and carefully proofread. Given the number of subjects students learn in school and the many teaching methods from which teachers can choose, it's easy to see that common agreement on an outcome can dissipate quickly when parents encounter teachers whose teaching approaches are not the ones they believe will help their children reach the stated goal.

Educators today also face another challenge to public schools as we have known them—one that goes well beyond differences in beliefs about teaching approaches and puts pressure on public schools to justify their continued existence. The debate about other options—charter schools and voucher plans—continues to intensify, and more and more parents are opting to educate their children at home. These factors, of course, challenge educators actually to improve public schools, but, even when particular schools are already doing a good job of teaching children, they must find effective ways to get the word out to the general public. If public schools are essential to the preservation of this democratic society, as many argue they are, then educators cannot continue to keep parents and the public at or outside the schoolhouse doors. They must instead invite them to come inside and work with them to educate "the living messages" who will be sent to the "world we will not see."

The public schools are the public's schools, but, even more important, the children in them are—and should be thought of—as the public's children.

BARRIERS TO INVOLVEMENT AND POSITIVE INTERACTION WITH SCHOOLS

Some parents want their kids to feel independent and they want to be a friend to their kids—and I don't think there's anything wrong with that—but teenagers need to have their parents involved in their day-to-day lives. We can't just send them off to school and expect four years later that they're going to be these great young adults. . . .

> *I don't worry about being "uncool," because there are too many kids falling through the cracks. I don't want to embarrass my sons; I understand peer pressure. But if I don't take the time to go to the school and find out what's going on, I can't expect somebody else to do that.*
>
> —Derrick Smith, a parent at James Madison Memorial High School, Madison, Wisconsin, "Staying Involved During the High School Years"

Involving parents (and community members) with schools is more difficult now than it once was. In the nineteenth century people were close to the schools in small towns where everyone knew everyone else. Teachers lived in the towns in which they taught. In the earliest days, teachers even lived with local families. There were many opportunities for both to interact informally and, even though there were conflicts between parents and educators, there was a much greater sense of shared purpose and common values than probably will ever be possible again.

At the beginning of the twentieth century, however, as many people left rural America for work in the cities and large numbers of immigrants came from abroad, schools became larger. Students who in former years might have learned with children of all ages in a one- or two-room schoolhouse were now divided into age-based grades and channeled into different tracks, depending on their future plans. These changes led to larger schools and layers of bureaucracy that were difficult for parents to negotiate.

Today, at the beginning of the twenty-first century, despite the efforts of many school reformers to create smaller schools by organizing "schools within schools" or "houses" so that no student will be anonymous, most schools remain large and impersonal. There are many reasons that parents hesitate to participate in or find themselves unable to be involved with their children's schools. And a growing number of parents has chosen to home school their children: The number of home-schooled children in one state was 25 percent higher in 2001 than four years earlier.[12]

One major factor is the demographic changes in society during the last century. Two-parent families are no longer the norm. In fact, even the definition of *family* is unclear because children may live with only a mom or a dad, grandparents, or guardians. For instance, a 1997 Census survey showed that 5.5 million children were being raised by grandparents.[13] Other children have two moms—or two dads. More women have full- or part-time jobs today than was true a few decades ago. Thus, work hours prevent people from volunteer work in the schools during the day and also make attending school events even at night more difficult. Because adults have less unstructured time and children often are involved in many activities, coordinating everyone's schedules is difficult. Before children are old enough to be on their own or drive themselves to various activities, a parent needs to find child care and serve as chauffeur. And all the work of maintaining a home and family, such as cooking, cleaning, or washing clothes, that many women used to do during the day still needs to be done albeit in less time.

For some families there may be additional powerful, personal barriers. Parents in the "sandwich generation" may face making a decision about putting an elderly parent in a nursing home or finding hospice care for one who is dying. Maybe the parent has a serious health problem. The first priority for parents living below the poverty line will be finding money to buy food and clothing. If the refrigerator is empty and the cupboards bare, they won't pay much attention to notices sent home from the school, and even the first grader's plea to *please* come to the class play will go unheard. There are only so many things a person can worry about and deal with at once, and family problems will always come first.

The level of parent involvement in public schools is relatively high, but not all parents participate equally. According to one report, in 1996 and in 1999, at least 90 percent of children had parents who participated in some activities, but (1) involvement is lower for children in higher grades, (2) involvement is related to household income and levels of parents' education, and (3) white students are more likely than black or Hispanic students to have parents who report participation in schools.[14] Some parents are hindered by their inability to speak English. Even when educators make attempts to communicate in other languages, they cannot be successful with all parents in a school if children come from homes where 37 different languages are spoken. Increasingly, such schools are common even in states typically seen as lacking diversity, such as Maine. Depending on their cultural or ethnic backgrounds, some parents may not see being involved with the school as appropriate. In some cultures, school and home should be entirely separate: It is not the parent's place to question what teachers do at school.

Other parents, especially those who are not white and middle class, may be intimidated by the school bureaucracy. Less educated parents may not understand the jargon-infested communications they get from the school. They often lack the "cultural capital," or know-how, that other parents have for negotiating with school officials to make sure their children are assigned to the classes of the teachers they choose. Middle-class parents know the value of and have provided the family support necessary for their children taking Advanced Placement courses. Less knowledgeable parents can do nothing—and may not even know—when their children are placed in dead-end general classes that will not prepare them for college. Educators may assume these parents do not care if children get postsecondary education, but the truth may be that parents without a high-school diploma themselves may really want children to go but have no clue about what it takes to get them there.

For parents whose own experiences were very painful, the school can be an unwelcoming or fearful place. If, as students, they received failing grades, got in trouble, and maybe were even encouraged by teachers and administrators to drop out before graduation, they will want to avoid anything that brings back bad memories they'd prefer to forget. So their children will be on their own. Sadly, history may repeat itself with the next generation.

After unsuccessfully trying to work with educators to solve problems with their children at school, some parents may just give up. They come to believe that nothing they can do will make a difference so they do nothing. Or, worse, their disillusionment with the school system may cause them to see all the children's future problems as the fault of the school—without realizing the damage they have done to their children. If children repeatedly hear parents criticize and blame the teachers, the children may develop the belief that everything bad that happens to them is "someone else's fault." They will never learn individual responsibility or acquire the confidence and competence needed to succeed in school or in life.

Parents, especially those of children who learn differently, may not be able to work cooperatively with educators because they make too great an emotional investment in their children. Having a child who has a learning disability or other problem that makes school learning difficult is not easy for the parent. Some parents believe that, once the child's problem has a label, there will be a solution. But humans are very complex organisms, and too little is known about the process of learning and its many variables.

Imagine, for example, being the parents of a nine-year-old boy who is diagnosed with dyslexia in grade 4. Your child already has a history of negative school experiences. As a result of the difficulty he had learning to read beginning in first grade, he may have been the object of other children's abuse (being called "dummy" or "retard," not being included in schoolyard play).

He probably always had poor school reports ("needs improvement" in many categories) and other experiences that make him believe he cannot learn. As a result, he may react by not doing as well as he could do on school tasks or may fail to do some of them at all. If he is big enough physically, he may become a bully who antagonizes others and starts fights at recess.

Even with special resources and assistance from grades 4 to 8, he may become an even greater problem at school once he reaches puberty. Now the hormones are raging along with all the negative feelings he has about himself and school. The principal calls to say that your son has threatened his English teacher. Meanwhile, you have had plenty of confrontations with the child at home. You are worried. You want your child to learn. You want him to have a good life. Can't the school do something? Shouldn't the school *have to* do something to help him? The answer, of course, is yes. His placement is changed, but the problems continue. You get more and more concerned, but nothing works. You are certain that there is more the school can do.

Your frustration with the school gradually shifts to anger with school officials for not fixing the problem. What they see as behavior problems that require discipline, you still view as stemming from his special needs. At this point your fears about his future may become so great that you are not be able to hear what educators try to explain. In fact, neither you nor the educators can say with any certainty what this child needs in the way of educational programming. There is no "right" answer anyone can give. Just when you as the parent need to be working with the school to figure out what else to try when the child's problem has become so difficult to understand, your emotional investment in this child prevents you from using reason to work with the educators in a thoughtful, collaborative way.

The irony of this sad situation is that both you and the educators really want to figure out what will help your son learn. But if things continue on this path, there will be no resolution, and your child will be the loser in a battle that did not have to be fought.

One way to prevent barriers, such as the lose-lose situation just described, is for everyone to consider any problem or disagreement from the other person's perspective. Teachers who imagine how they would feel if they were the parents will be more sensitive to the concerns parents have about an individual child. Parents who try to understand the roles and responsibilities teachers and administrators have for educating both the individual child and all other children will respond more reasonably to suggestions with which they initially disagree.

Building better relationships with each other, although in the best interests of children, makes some parents and some educators anxious for many of the reasons previously discussed. Noted author and scholar Andy Hargreaves, who argues for building "a social movement for educational change," powerfully explains the situation facing everyone who cares about educating the nation's children. Although he addresses educators, his words are important for everyone:

> We are now at a crossroads in the future of teaching and public education. One road leads to the creation of exciting and positive new partnerships with groups beyond the school. . . . The other road heads toward the deprofessionalization of teaching as teachers crumble under multiple pressures, reduced time to learn from colleagues, and derisive discourses that shame them for their shortcomings, blame them for any number of purported failures, and sap their spirits. . . .
>
> The road we take should not be left either to fate or inertia. It should be shaped by the active intervention of all educators and others in a social movement for educational change that advances the principle that, if we want better classroom learning for students, we have to create superb professional learning and working conditions for those who teach them. Perhaps this is the most urgent reason why teachers must overcome the immense and understandable anxieties that many now feel about opening up their practice to parents. Closing the door on parents and trying to control the interaction with them is in the worst interest of teachers.[15]

The social movement that Andy Hargreaves calls for begins with rethinking what it takes to educate all children well. The next chapter addresses other important differences in perspectives about schools and the limitations and tensions of the traditional model of the home-school-community relationship.

2

THE TRADITIONAL
HOME–SCHOOL–COMMUNITY RELATIONSHIP:
TENSIONS AND LIMITATIONS

The history of American education shows that the relationship between parents and educators has often been prickly and problematic—and it is likely to continue to be so in the future. Although parents and educators share the same goal—that all children will learn and be successful in school—they bring different perspectives to this challenge. And because public funds support the schools and businesses hire their graduates, the opinions of people who do not have children in school also matter when decisions are made about public education.

Despite the fact that parents, educators, and community members have a vested interest in finding ways to educate all children well, the communities in which they work closely together have been the exception rather than the rule. This chapter, which examines the tensions and limitations of the home, school, and community operating as separate satellites in children's lives, underscores the need for rethinking the traditional way we think of the home–school–community relationship.

UNDERSTANDING DIFFERENCES IN PERSPECTIVES

Several factors account for differences in perspectives, not just between parents and educators but also among parents and among educators. Differences can create tension, but the most powerful one can be described as a dilemma, "my child" vs. "all children."

"My Child" vs. "All Children"

> *I've said to teachers, "You have my child for one year. The child I'll have until the day I die. . . . [T]here are things I know about my child that you don't, simply because I've known my child longer than you have. . . . [W]hen my child leaves your classroom and moves on to someone else, she's still my child. You have some feelings about her . . . but she's still mine. . . . I'm the consistent part of this child's life.*
>
> —Parent, quoted in Linda Ginn, "Following Their Kids to School: A Study of Parental Involvement"

> *I don't give a tinker's damn what parents think! That's the problem with asking parents for their input. They think that we will use it all—when they're just thinking about what's good for their kid. We have to think about what's good for all kids.*
>
> —Teacher, quoted in Jean L. Konzal, "Our Changing Town, Our Changing School: Is Common Ground Possible?"

Parents will always be more concerned about their own individual children, "my child," than teachers will be. And even as teachers and administrators attempt to resolve a situation with only one child, they always have to consider this student's needs in the context of the needs of all children. For instance, when one student makes a special request, the teacher often replies, "But I couldn't do that for everyone, so how can I do that for you?" Teachers must enforce the classroom and school rules that govern the behavior of all students. While some recognize that interpreting the rules differently for different children may be fairer than interpreting them uniformly, many worry that chaos will result if they make exceptions for any individual.

On the other hand, parents have no qualms about insisting that their children be the exceptions. For example, when a preteen uses the argument that everyone else is wearing lipstick or going to the mall, the argument is likely to fall on deaf ears: "If everyone wanted to jump off the bridge, should I let you jump, too?" The parents' focus always will be on what they think is best for the individual child. And yet, ironically, sometimes parents seek help from the schools to lay down the rules. For instance, one principal told the story of a parent who didn't want her child to drive to school. She asked the principal to tell the child that he couldn't. Of course, since there were no school policies prohibiting this, the principal couldn't comply with this parent's request.

Another telling example of parents' focus on "my child" was the response a former superintendent of schools gave to the question "Do you prefer homogeneous (tracked) or heterogeneous (untracked) classes?" He had been an advocate for introducing untracked classes at the middle and high schools in his district. It was easy to see that the question was a tough one for him to answer because he verbally danced around it for a moment or two. When pressed by the interviewer for a response, he finally said, "I think if I were really altruistic, I would say, yeah, kids ought to be heterogeneously grouped and teachers ought to be trained to deal with those kinds of issues . . . [but] ideally, if I had it to do all over again, I would insist that Ronnie [his son] be with nothing but top kids." When there was a conflict between his professional knowledge and his parental perspective, the superintendent, as any parent would, put what he considered to be his child's best interests above those of other children.

Parents as Active Participants vs. Teachers as Professionals

> *I am your worst nightmare. I am the parent who will make phone calls, write letters, and give speeches at school board meetings if I think that your great new educational idea is nothing more than the reform* du jour. *You and I see change from different perspectives.*
>
> —Tim Zukas, "An Open Letter from a Parent"

This dilemma, "my child vs. all children," is not the only difference between parents and educators. Another twenty-first-century reality is the paradox that, at the same time school reformers argue for more parent participation in schools, they also advocate a more professional role for teachers. So, as parents become more involved in schools and try to participate in making decisions about children's education, they encounter professional educators, who believe that they have specialized knowledge that others do not have and should have more autonomy for making decisions about what and how they will teach students in their own classroom. One teacher expressed it this way:

It would be fairly rare in modern medicine for doctors to bring their patients together and ask them which kind of technical technique or chemicals would best help the healing process. Now it may be worthwhile to bring patients together about service issues and fee issues and how comfortable they feel with the doctor... but the technical aspects are left to the doctors. Why is education different? Because most people in our society have a high school education, there's an assumption that most people are educational experts or at least they're close to it, even though that isn't the case."[1]

It doesn't take a rocket scientist to see that these two seemingly contradictory aspects of current school reform can generate conflicts.

Philosophical Differences

Philosophical differences create tension not only between parents and educators but also among parents and among educators. Based on their past school experiences and knowledge about teaching and learning gained over many years, people develop differing mental models, ideas of what a "good" school should be. Although people are often unaware of these powerful, personal mental images, these often determine their response to a particular practice. For instance, if spelling tests were emphasized when they went to school and their teachers always insisted on correct spelling, parents are likely to view a teacher who tells their children not to worry about spelling on early drafts of a writing assignment as misguided or incompetent.

Because there is no one mental model for parents and another for educators, both groups represent a variety of philosophical perspectives. Thus, reaching consensus on any school issue is difficult—and sometimes impossible. The continuing debate about how best to teach children to read is a good example. Some parents and some teachers believe the phonics approach (children learn to read by focusing on learning the sounds that comprise words, by sounding out words, and often by circling the correct letters and blends on worksheets) is the "right" way. Other parents and educators, though, are just as certain that the whole language approach (children learn to read by choosing books that interest them, using a variety of strategies for figuring out unfamiliar words, including phonics, and writing their own stories, often using "invented spelling") is best. So, in one elementary school, it is possible to find teachers using both approaches and parents asking that their children be assigned to teachers whose philosophy mirrors their own.

People's beliefs are powerful influences on their thinking, and they do not change these beliefs easily. Educational philosopher Thomas Green, for instance, says that people hold to beliefs because they "see there are good reasons for them, and . . . they tend to hold to them as long as they do not discover better reasons or better evidence for some other belief."[2] Researcher Frank Pajares points out that when people get new information that contradicts their existing beliefs, they may distort it or reject it rather than change their beliefs. "[B]elief change is the last alternative."[3] According to another scholar, Milton Rokeach, the earliest beliefs people have are the most central and the most difficult to give up, but people also vary in their willingness to consider adopting new beliefs: Open-minded people have relatively open belief systems while the opposite is true for close-minded individuals.[4] His work explains why parents' and teachers' own school experiences create the mental lenses through which they view classroom practices and why some people are less likely than others to adopt different mental models of how school should be.

Yet if researcher Paul Ernest is right, change may be possible when people are introduced to new knowledge. Even though belief systems are not the same as knowledge systems, he points out that beliefs also have a "slender but significant knowledge component."[5] Rokeach also thinks that beliefs have a cognitive component, noting that "people vary in their relative knowledge about things believed and disbelieved. In some people the discrepancy in knowledge is great, in others less great."[6] One study of parents' beliefs about preferred classroom practices does suggest that the parents with the most knowledge of and experience with different methods of teaching are more likely than others to prefer nontraditional practices.[7]

Sometimes, as a result of school-change efforts, teachers gain new knowledge that leads them to develop a different philosophical perspective. For instance, the school district may provide in-service and professional development opportunities to help teachers learn new ways of thinking about teaching and learning. They participate in ongoing study groups and discussions with each other and with colleagues in their school and across the nation about how children learn and how teachers teach. On teacher-workshop days teachers engage in activities or work with outside consultants who help them develop skills in using different teaching approaches. As a result of this new knowledge, teachers see that practices different from their current methods have value. Gradually they change their mental models of a "good" school.

Parents, however, who have not had such learning opportunities, continue to judge school practices with their old mental models. Because they are not ready to accept the change when the school introduces it in their children's classrooms, conflict between educators and parents is likely. As one teacher said, "Sometimes, parents don't understand the professional part of it. Parents aren't involved in the national standards in math and the National Council of Math Teachers. These professional groups want students to do real world math problems instead of just learning the times tables. And parents hadn't been part of that conversation, that's why they were so opposed to it. Their biggest comment was, 'When I went to school I did it this way.'"[8]

Parents and Teachers vs. the Bureaucracy

> *[What happened to my son] illustrates the difficulty too many parents have in making themselves heard above the groaning of a constipated educational bureaucracy. These are challenging times in which to raise children, and parents should at least be able to count on schools as their allies.*
>
> —Howard Good, parent and former school board member, "One Big Family"

The bureaucratic nature of schools prevents parents and teachers from productively working together. Educational historian William Cutler argues that bureaucracy, which came to schools in the beginning of the twentieth century, "did not drive parents and teachers completely apart." Although his description suggests a benefit of bureaucratization for parent involvement, the role parents played then—one that continues today—may explain why parent participation is so limited:

> The professional identity and occupational opportunity that it [bureaucratization] fostered gradually gave many teachers, principals, and superintendents the self-confidence to reconsider the home-school relationship. It did not have to be as antagonistic as the first public officials frequently thought. Bureaucratic reform led educators to contemplate how parents could be transformed from vocal adversaries to loyal advocates by building them into the school's organizational framework. Properly sorted and arranged, mother and fathers could be an integral and valuable part of the American educational system.[9]

Parents may resent being "properly sorted and arranged"! Yet the truth is that, even today, parent access in traditional schools has been limited and controlled by educators in ways that are not helpful to students. But teachers, too, are constrained by a rigid bureaucratic organizational culture.

While schools and businesses first adopted bureaucratic frameworks and procedures to improve their *efficiency*, schools have been slower than their corporate counterparts to consider whether this management approach improves their *effectiveness*. Paul Houston, executive director of the American Association of School Administrators, notes that in the past, a successful superintendent had to be good at the "killer B's . . . things like buildings, buses, books, budgets, and bonds."[10] The good superintendent in a hierarchical organizational structure is a good manager. The problem is that, even though people have specified roles and expectations are clear, the power to make decisions is usually in the hands of the few people at the very top. Teachers have little or no power to affect what happens, and parents are not even included. The larger the organization, the more complex—and time consuming—the process will be for getting anything done. Only the bravest parents even attempt to negotiate the maze and then only when they have what they consider to be a serious problem with a child.

In some ways the bureaucratic structure serves to protect the school and its teachers from outside pressures. Many principals, for instance, have been socialized to be gatekeepers or buffers, who constantly guard the walls of the institution in an almost medieval way. Their job is making sure that outsiders, parents or the public, do not interfere with the routine functioning of the school. Given the number of recent incidents of school violence, principals have a good reason to be concerned about the safety of students and staff, and controlling access to the building is essential in this regard. But, security concerns aside, this bureaucratic sensibility also means that parents and teachers interact in formal and ritualistic ways, preventing them from having the kinds of informal interactions necessary for building trusting relationships.

Perhaps the idea of the school as a medieval fortress is a bit extreme, but, as the next section shows, the existing model of the home–school relationship has several built-in problems that limit—or prevent—a collaborative relationship between parents and educators.

THE OLD PARADIGM: HOME, SCHOOL, AND COMMUNITY AS INDEPENDENT SATELLITES

Educators ask: What can parents, community members, and organizations do for us?

Traditionally the school, the home, and the community have been seen as related but separate. The relationship among the three can be described as

Figure 2.1

Old Paradigm

Home, School, and Community as Satellites
Separate, Independent

- Focus on schooling (academic skills and knowledge)
- Relationship school-controlled
- Bureaucratic, impersonal, one-way communication
- Self-protective, defensive
- Hierarchical, not everyone included equally
- Cultural and social differences ignored or erased; some families and students marginalized
- Preserving power
- Parents as problems or critics
- Community "out there" except when needed or adversarial

Educators ask: What can parents, community members, and organizations do for us?

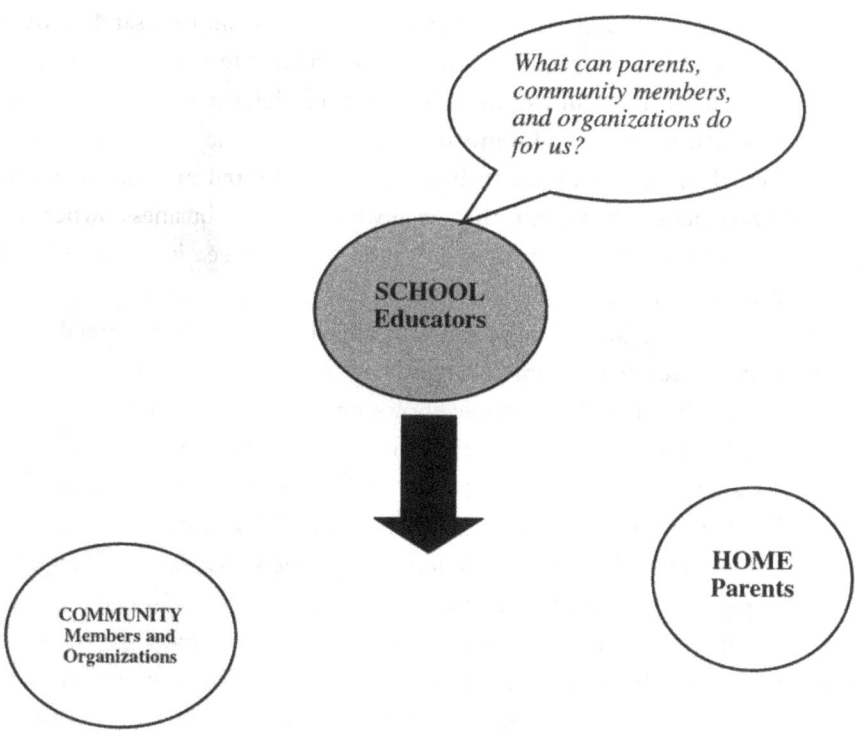

"satellitic" because each operates as a separate satellite within its own orbit in children's lives. Occasionally, when there is a need or a contact cannot be avoided, people will loosely link one to another—home to school, school to community. Usually, however, like having several documents open at once in a word-processing program, only one is in the foreground; the others are hidden from view.

Even someone assessing the satellitic model through a bureaucratic lens might find it problematic because it lacks an overall hierarchical organization that encompasses all three. Thus, this model is not efficient because, without any specific roles and responsibilities assigned to each sphere, or an organizational chart that defines who reports to whom, educators, parents, and community members and organizations can, without knowing it, duplicate their efforts. Or, more problematic, they can be working at cross purposes.

In each separate sphere or satellite, educators or social workers do their work and parents raise their children as best they can. But people tend to operate out of their own self-interests. They do what they think is best, trying to preserve what they believe is theirs and advocate for what they think they or their students/children need. When someone from another satellite questions them, as, for instance, when a parent objects to a teacher's means of disciplining a child, they can't seem to help getting defensive. An intruder has come into their territory and wants to take something they believe is theirs.

When their opportunities to interact are so limited and almost always formal, the teacher, the parent, the social worker, or the business owner cannot develop personal relationships. Instead, what they see is the other's role or position rather than the person who's in it. Many of us can remember being surprised the first time we saw a teacher buying milk and bread at the grocery store. Teachers shop? Teachers eat? In retrospect, such a memory makes us chuckle, but this example shows how easy it is to view individuals not as whole persons but only in the context of the roles in which we know them. In addition, roles remain rigidly defined with impermeable boundaries. Teachers (especially at the secondary school level) may see their role solely as focusing on developing students' academic skills and knowledge and resist suggestions that they are responsible also for the child's social and emotional development. Because people's individual perspectives are narrow—and because they have little opportunity to interact with people from other spheres, their relationships with anyone in another of the three satellites (home, school, or community) tend to remain relatively static and impersonal. In addition, people develop a language that works well to communicate with those within their sphere but has no meaning for those who are outside. With no understanding or recognition of others' perspectives, working together can be difficult or elusive. Rather than inviting parents to work with them, educators too often send the message to parents through their actions and attitudes: "Here's what *you* need to do to help *us*." Parents may have an important role to play in children's education, but it is subordinate to the role of the professional educators.

The discussion that follows explains other limitations of the satellitic model, in which school is viewed as an institution for schooling. When their mission is to provide students with academic knowledge and skills, schools can become more efficient at schooling, but they are unlikely to be very effective in educating the nation's children. Educating children means much more than stuffing their heads with knowledge and information. Educators alone cannot help children develop intellectually, personally, socially, and morally—develop all the knowledge, attitudes, and skills they will need to be productive citizens and caring people as adults. Educating children well requires contributions and commitments from everyone in the community.

Educating children in this way is a goal everyone would support, but, as long as the old paradigm persists, the traditional way of doing things suggests that we will never achieve this goal of educating all children well.

Families and Schools

Parents are children's first and most important teachers. Because they teach children what they believe is important for them to learn, children in different homes acquire different knowledge, values, and attitudes. What educational historian Lawrence Cremin calls "the family curriculum" differs on the basis of race, class, ethnicity, sexual orientation, or other factors, such as the educational background and prior experiences of the parents.[11] The more educators know about children's families, the more effectively they will be able to teach them at school.

Although elementary and special-education teachers know much more about the families of their students than do secondary school teachers, few teachers today know enough. Most teachers, for instance, do not make home visits; they may not be fluent in the parents' language; and they may have almost no contact with parents who, for any of the several reasons discussed previously, may be reluctant to be involved with the school. Yet what teachers and parents do not know about each other can create misunderstanding

> **Sidebar 2.1**
>
> *Marilyn Brown*[*]
> *Elementary teacher and parent*
>
> ### Schools Make Unreasonable Demands on Some Families
>
> The first time I held my own firstborn, I knew that the love I had for this child was the same kind of love the parents of the children entrusted to me have for their kids. I felt a new heightened awareness of responsibility since I am given the privilege of spending a good part of my day with parents' most precious gifts. I also now know firsthand how difficult parenting can be. As my husband and I juggle the needs of our three sons and the demands of our jobs, we have to make a conscious effort to make sure we also have quality family time. It takes a huge amount of effort and know-how to sit down as a family, have a discussion, and eat a healthy meal together. Some nights it is not easy. Those are the times when I realize that schools make unrealistic demands on some families.
>
> We expect students to do their homework, but what about those assignments that require parental assistance? I still have nightmarish thoughts about the first-grade "snake report with illustrations and a diorama." Whose project was that anyway? What first grader could do a full-blown research project with a bibliography? Did we do it? Of course, and we got a check ++ with smiley faces. I was thrilled when we were finished with the project and, as a parent, proud that my child felt some ownership, didn't cry too much, and actually was able to read the report. Of course, the educator in me wondered what would happen to the kids who didn't have the family support to accomplish the task. The teacher, I suspect, felt this was a great project since it went along with all the latest educational trends. It was a "performance task" (actually multiple tasks!), and she had sent home a nicely done rubric to help guide us through the task(s) components and criteria.

and/or conflict. And rarely does anyone discuss the negative effects of the "hidden curriculum" in most schools, the purpose of which is to socialize children into a white, middle-class value system. These values may be in opposition to those the family holds dear.

The differing family curricula students have experienced before they even get to school mark some students for success while others are labeled and often live out their school lives as failures. Researcher Shirley Brice Heath, for example, found that young children in white working-class and black working-class families develop language patterns and practices at home

---

[*] The only contributor who requested we use a pseudonym.

that conflict with those of the school. Conversely, children in mainstream, school-oriented families already have a great deal of practice at home doing what they will do more of in school.[12] Other children, however, will have great difficulty adapting to the demands of school.

When children come from nonwhite, non–middle-class families, they often cannot find themselves or their families in the official school curriculum. The social studies curriculum that tells the story of the United States from the perspectives of the white Europeans who settled here, for example, sends negative messages to Asian, African American, and Native American children. Their ancestors were either the "bad guys" or they didn't exist at all. When Franco-American children were punished in school for speaking their home language, as they used to be, school taught them to be ashamed of their family backgrounds. This practice continues today, though in more subtle ways, for thousands of children whose first language is Spanish. The more "traditional" the school curriculum is, the more damaging it will be for some children and their parents.

In addition to transmitting culture, values, and beliefs to their children, parents get involved in teaching the school curriculum at home, as, for example, when they attempt to help children with schoolwork. Often this support is helpful for the children and their teachers, but that is not always the case. What the family does at home may be in direct conflict with what teachers are doing at school, confusing children and perhaps forcing them to "choose sides." For example, attempts by educators to help parents understand new methods of instruction by getting parents to help children with their schoolwork may hinder more than it helps. When parents do not understand the methods teachers use, they can confuse their children. Meetings to explain new methods to parents may not be effective—or even a good idea—because, as two researchers discovered, some parents found what was covered in these meetings to be "over their head." At the secondary level, parents may not be able to help because their children have already passed them in knowledge.[13] Other researchers point out that "[e]ven relatively well-educated parents felt at a loss with modern teaching techniques. As one father put it, even his elder son who had just left school could not help his younger daughter with her math because things had changed so much within that period of time."[14] Immigrant parents may resist the school's efforts to involve them in any way because, in addition to their lack of fluency in English, many of these parents may expect that it is the school's job to teach, not theirs.[15]

Studies have shown that when parents understand and support what goes on in schools and classrooms, children are more successful.[16] Many times, however, parents don't support a school's practices. As a result of school reform, many teachers now have less-structured classrooms than was true in the past. Students work on lessons with classmates, and the relationship between the teacher and students can be quite informal. Some parents may be more accepting of these changes than others. One study suggests that parents with low status in our society—working class, poor, immigrants, African Americans, Native Americans—for example, tend to prefer rules and structure and to depend on the security of past practice and tradition to guide their conduct while high-status parents—white middle class—are likely to be more flexible, dynamic, and adaptable.[17] Other studies suggest that progressive styles of teaching are more likely to be favored by highly educated professional parents and opposed by working-class parents.[18] However, more recently, high-status parents in some communities have opposed the introduction of progressive practices, such as untracked classes, because they fear the changes will prevent their children from getting into good colleges; this outcome puts some of the earlier findings into question.[19] Thus, educators contemplating changes in schools need to consider the multiple perspectives through which any group of parents is likely to view the innovation.

Socioeconomic and cultural differences affect parents' responses to classroom and school practices. In some cultures, for instance, parents do not want their daughters pairing up with boys to do an assignment—important information for teachers who want to incorporate cooperative learning—or staying after school to participate in extracurricular activities.[20] When elementary teachers adopted an "open-classroom model," encouraged children to use their first names, allowed the use of black dialect in school, and instituted other progressive practices, parents saw no benefits for these changes. Instead, "[a]ll they saw when they looked into classrooms was chaos, sloppiness, and uncivilized behavior . . . it all looked like 'play,' not work."[21] Likewise, researchers Michael Young and Patrick McGeeney report that some immigrant parents also "did not appreciate the function of play in learning nor could they understand the refusal of the school to give children homework, particularly the formal rote-learning sort which they felt would have enabled them to follow the progress of their children."[22] A working-class parent in another study was upset when the school made changes because he felt "progressive styles of teaching were experimental, and . . . children had lost ground by being the unwitting subjects of an experiment."[23]

When teachers think about parents, they see them as helpful when parents reinforce what they are trying to do at school. They don't want parents telling them how or what to teach. Teachers want parents to make sure kids come to school, rested, fed, and with their homework done. The home and the school are separate but not equal, and what parents do with their children at home should complement rather than contradict what teachers do. It is easy to see how this model of the family-school connection disadvantages many children and their families.

Home–School Communication

Positive and ongoing communication is central for building relationships between educators and parents, but too often the traditional means of doing so is one way. Parents, especially at the elementary level, can be overwhelmed with information from the school. Yet what they get are notices about upcoming events or program changes, forms to sign, requests for information the school needs, progress reports, or notes from the teachers about a problem with the child. Although some schools realize the need for sending information written in Spanish as well as English, educators may not even think about the parents who speak other languages or may be illiterate, even in English.

Sometimes the information sent home is very important—something parents need to know—but, when a message contains too much jargon, it might as well have been written in a foreign language. How many parents, for instance, would understand what the following information actually means for their children?

> The school board recently approved the middle school's faculty request to introduce heterogeneously grouped, team-taught classes in the major subject areas next September. Teachers will attend a summer workshop to learn methods for individualized and collaborative learning as well as how to develop student-centered interdisciplinary units. They will also spend at least one day creating rubrics for assessing student work in line with the new state learning standards. We are very excited about these changes and how they will help students. We expect that when the changes are fully implemented, you will see how much they have benefited your child.

Even a college graduate might have trouble understanding the "educationalese," but which parents would call the school to ask for clarification?

> **Sidebar 2.2**
>
> *Marilyn Brown*
> *Elementary teacher and parent*
>
> ## It All Comes Down to Communication
>
> I will never forget the first parent-teacher conference I attended when my son was in kindergarten. He seemed to be doing fine academically and socially, but he didn't have a check mark next to the box for "Able to tie own shoes." I knew my child could tie his own shoes, but, trying to be helpful, I had bought him sneakers with Velcro closures. Because he couldn't tie his shoes quickly, I figured there would be one less kid the teacher would have to worry about. Had I known that "tying shoes" was on the report card, I would have made sure he had shoes to tie! Needless to say, when my middle son went to kindergarten this year, I made sure he had sneakers with ties. He desperately wanted a pair of shoes from L.L. Bean Kids that were slip-ons, but I told him he could have them only if he promised that on "gym day" he tied his own shoes. I didn't want there to be any question about his ability. This is a true of example of the importance of letting parents know classroom benchmarks and expectations up front.
>
> My son's second-grade teacher was a true gem of an educator. She truly knew the meaning of *communication*. She did not feel comfortable with e-mail but was always available in person or by phone. I will never forget the first time the phone rang and I heard her voice. Being conditioned to believe that teachers call only when there is a problem, I immediately asked, "What's the matter?" Come to find out, there was nothing troubling. The teacher simply wanted to know how we thought things were going. This was a wonderful opportunity for me to share my concerns about my son's reading fluency. Throughout the year she called on at least four other occasions. When I talked with other parents, I found out that this teacher calls everyone this often. I quickly did some mental math calculations to figure out that she made approximately 92 parent phone calls on her own time! I applaud her dedication.

No one wants to look stupid. The school that sends such messages home to parents has failed to heed what many writing teachers emphasize with students: the importance of knowing your audience and writing so that they will understand what you want to communicate.

Parents in Schools

Parents—though not all parents—are involved in schools in several ways: by attending activities, athletic competitions, and parent-teacher organization meetings; by volunteering as room parents or library assistants; and, less often, as members of school advisory councils. In all instances, however, their

Sidebar 2.3

Betsy Moyer, parent
Brunswick, Maine

Advocating for Children with Special Needs: A Parent's Perspective

Our experience with our children in the school system has mostly been positive. The teachers have been great. It has also helped that we have advocated for our children and the teachers have been very receptive. As parents we need to be on top of what is going on. Communication is a big plus. If you can establish a good relationship in the beginning, it paves the way.

Having a child with special needs has not been easy. Finding the resources and help that you need can be difficult. Having a child who was very active and being told that my child was an overactive child and would grow out of it by a doctor was difficult especially knowing my child. I did not accept that answer. I took it upon myself to get the answers I needed which took a lot of work. Asking friends, teachers, other medical personnel and whoever it took to point me in the direction I needed to go to get the testing I needed to find out what was wrong, was a very hard thing to do. People who are trained to diagnose and test kids are very knowledgeable of the laws. They tell you what the schools need to provide for your children and they are willing to work with you in getting what it takes for your child to succeed.

Getting into third grade and higher was difficult. This was when the kids start to recognize who's different and who's not and they also start to form little cliques. One of my pet peeves is the amount of bullying that goes on at such a young age, which is really scary. My children have been subject to that and it has gotten the best of their self-esteem. It appears that schools are addressing it but it needs to start a lot sooner.

Being open and up front with educators has a positive outcome. In the beginning of the school year I try to meet with teachers, resource people, ed techs and even the principal to let them know problems, issues or what has worked and what has not worked. The teachers have been very thankful especially when they don't know the child and knowing that there will be 25 to 30 other children in the class.

Mentoring is becoming a strong point in schools. It's great to have peers that you can talk to and share things with that you wouldn't share with your parents. They also help with class work and homework. The ones we've had have been high school seniors and college kids. It's been great because it takes some of the pressure off the teachers.

I feel bad for teachers: They have so many demands on them, the class sizes are large and they don't get to give the kids a lot of one on one time and they can't teach properly.

We have found in junior high that the teachers who are fairly young can really relate to the kids and can teach them in a way that is fun and that they also understand it along with being educational. The projects my kids have done have

(continues)

> **Sidebar 2.3** *(continued)*
>
> been awesome and yet fun to do and they learned a lot from doing them. They felt good about themselves after doing them.
>
> But one teacher in junior high likes to threaten students with detention and cannot give a student any positive feedback. I thought threatening wasn't allowed. If a student threatens they get detention, suspended, or even expelled. If a teacher does it it seems to be okay. Something doesn't seem right. We addressed the problem with the teacher but nothing happened. I tried to pull my child out of the class but was unable to do that, so I told my child he did not have to do any work in that class and if he got threatened with detention he was not staying and if he flunked that was not okay either. We as parents stood up for our son and had people backing us except for the special-ed director and superintendent. Any ideas on how to beat the system would be greatly appreciated.

influence on school policy and programs is limited. And the parents with whom educators have the most contact are usually well-educated, professional people with status in the community.

Schools typically do a good job of inviting parents to attend events at school, but these events may not be helpful in building relationships between parents and educators. Parents attend school events that are open to the general public, such as athletic contests and art exhibits, but their role is as spectators, not active participants. Other events are primarily informational. Parents of ninth graders, for instance, are told what to expect in the fall when their children enter high school. Or the principal and faculty members may schedule an open meeting to explain a change, such as the introduction of block scheduling, which is already a *fait accompli*. They may welcome questions from parents about the change, but they will discourage any criticism. As with written communication, if too much jargon is used in the oral explanation, some parents may have no clue what the new program means for their children but will be too embarrassed to ask. Only the most courageous people usually feel comfortable speaking in a large group. One white, middle-class, well-educated father put it this way:

> Non-teaching parents are hard put to maintain their enthusiasm within an atmosphere of "educator-speak" and I have found this to be true in my case. Parent involvement in the reform effort has dropped out to an alarming amount. Professional jargon should be a device to speed communication only within members of that profession. It has no use within a mixed group

where non-teaching parents are reluctant to admit ignorance on so many of the terms used by educators. These professional power words are very very uncomfortable when parents are involved in the process. Lots of people have a huge problem with them. They don't want to put up their hand and say, "I don't know what you said." And if you're not careful you can get behind in that understanding. And after a few meetings they say, "I don't really know what's going on." And so the parents drop out.[24]

Parent conferences and open houses are the most typical ways parents can find out about their children's schoolwork and meet their teachers. Yet even in these instances, their participation may be too limited to be of much benefit. The high-school open house, for example, often is organized so that parents go through an abbreviated schedule of the child's day. During the ten-minute class sessions, the teachers do all the talking. There is not enough time for questions, and parents can't ask questions privately afterward because parents for the next session are already coming in. While these programs are not designed for parents to have one-on-one conversations with teachers about their specific child's needs, parents may not understand this and become angry when, instead of specifics about "my child," all they can get is general information.

Also, parents with more than one child in a school have the problem of trying to go through the schedules of two or more children at once. Single parents can't meet all of the teachers so they have to choose. For many parents, the evening can be chaotic and stressful as they hurry from one session to the next, uncertain where the various rooms are located in the large building. Instead of gaining useful knowledge to help their children, they get pounding headaches. Open houses may be useful, but they aren't enough for parents who want personal information about their children.

Educators, of course, recognize that parents need opportunities to discuss their own children's progress so most schools set aside time at least once a year for these. Yet even when parent conferences are scheduled in the late afternoon and evening, some parents cannot attend. And at the secondary level, since many high school teachers see more than 100 students each week, it can be a good thing for the school if some parents don't come. The history teacher who teaches that many students can't give more than a few minutes to any one parent—not enough time to have a meaningful discussion about a child's work. If, as is typical, teachers control the conferences, spending most of the time talking *at* the parent, parents not only don't get the information they feel they need, they also may be frustrated because they never get a chance to say what was on their minds or ask a question.

Parent volunteers are common especially in elementary schools, and educators both welcome and depend on such help. Volunteering is good for parents who have the time and inclination to be involved with the school in this way. They can get to know the teachers on a personal level, and, because they are able to see classes in session, they can develop a good understanding of the school's culture, programs, and teaching methods. And teachers can get to know these parents well. As one savvy parent said, "I made it my business to be seen in the school because I know that when teachers know parents, when they see them in school, they will take special care to treat their children well." Once again, though, only some parents become volunteers. Parents who work don't have the time, and others wouldn't consider coming into the school, perhaps because of cultural differences, language barriers, or low levels of self-esteem.

In addition, educators do not see parent volunteers as partners in the education of children. Instead, they depend on them to raise money for the band and athletic teams or to buy new computers, to chaperone field trips and school dances, or to serve as room mothers who will call other parents to bake cookies for the bake sale. Teachers and administrators expect volunteers to provide support and services, not suggestions about other ways to run the school or teach the students. As subordinates to the teachers, parents work *for* rather than *with* the school.

Parents are important contributors to students' learning, but educators also see this function as kind of a one-way street. However, in addition to making sure children do their homework and assisting them in completing school assignments, parents can encourage them to read books at home, take them to museums, use vacation trips as educational opportunities, involve them in construction jobs around the house, and myriad other activities that contribute to their child's education. Many educators typically limit their thinking about what parents can do at home to help children learn to those activities that directly serve or support what teachers are doing in school. Educational historian William Cutler sees this situation in a more complex way: parents as outsiders who are controlled by teachers, the insiders:

> Shaped by the politics of interest, the relationship between the home and the school has entailed considerable ambiguity and possessed a sizable potential for conflicts born not so much of fundamental differences as misun-

derstanding. Both parents and teachers have often misinterpreted what the other expected them to do.... The status of parents turns out to have been even more enigmatic, for once bureaucratization occurred, they qualified as both insiders and outsiders at school.... Classifying parents as outsiders, education invited mothers and fathers to get involved with homework or special events without giving them any authority or even making serious work for them to do.[25]

Although administrators may resist the idea of involving parents as advisors and may even put up barriers, many have recognized the political good will they can gain from involving parents in limited ways. Very few administrators, however, have seriously considered inviting parents as partners in decision making. Public meetings about school programs or policies are primarily informational. Parents listen to what educators have to say with little or no time allowed for their questions or comments. If the school has an advisory council, the parents selected to participate are often the same ones who tend to fill the volunteer slots; they are usually white and middle class. They can't serve as good representatives for all parents because they are not likely to know much about the perspectives of parents from lower social classes, other cultures, or different races. Principals often invite the parents who are most powerful in the community to serve on an advisory council. Typically it will function as a focus group—a means for administrators to check the pulse of the part of the public whose support they need—rather than as decision-making body.

There are times when administrators *must* include parents because they have no choice. Legislation in some states, for instance, mandates parent participation in developing school improvement plans; the rules of accreditation agencies also require parent membership on committees that write various aspects of the school's self-study. Yet in these cases, parent representatives are usually outnumbered by educators. For example, the membership of the committee in one school charged with writing the self-study section on the school's philosophy was comprised of one parent and six teachers. Because this ratio of parent to teachers is more the rule than the exception, anyone can see that the educators—not the parents—will control what will be decided.

Too few educators think of parents as learning partners even though research suggests that this role is critical to the success of many school reform efforts.[26] When parents do not understand the benefits of a particular program or teaching method for their children, they will not support it—and, in some instances, they will rally the support of other parents to change it. These efforts can grow into full-scale community conflicts, which are painful for everyone involved. Yet, typically, when schools do plan events to help parents learn, they choose topics that educators think will improve parenting skills. These topics, such as "Talking to Teens about Drugs" or "Helping with Homework," are useful and necessary—and will help children. The problem is that, without parents' input, educators make the decisions about what they need to learn. They don't even consider the idea of parents and teachers learning together. They fail to see how parents and teachers could plan a workshop in which both groups, at the same time, could develop new understandings about an innovative practice the school plans to introduce, such as writing as a tool for learning or creating interdisciplinary courses. And perhaps most telling, they do not even entertain the idea that they might learn something valuable from parents.

This outdated model of the school keeps power in the hands of educators. Educators decide how and when to invite parents in and control those interactions in ways that limit parent interference. Educators, who have professional knowledge that parents lack, determine what and how students should learn. Parents are important and should be involved, but their main role is supporting what the school does. This satellitic model keeps home and school separate but not equal. Because educators view their main responsibility as fostering students' academic development, they tend not to think about the other aspects of children and their lives that influence their progress in schools. And both parents and educators fail to see another important link in the education of children: the community.

Parents, Educators, and the Sometimes-Adversarial Community "Out There"

The satellitic model reveals that the community is a third sphere, distinct and separate from the home and the school. Within any community are many untapped resources, which, if more wisely used, could assist parents and educators in helping children grow and learn. Social service agencies, businesspeople, volunteer or service groups (Big Brother/Big Sister, Kiwanis, Rotary), youth groups (Scouts, 4-H), senior citizens, and other community members provide support for families or children. Because there is no structure for ongoing communication and collaboration, people don't always know what is being done or how an individual or a group might help solve a particular problem. A high school administrator shared a telling example of this disconnection:

> Part of my job is dealing with the attendance for the thousand kids in this school—a thankless job because no one seems too concerned about whether kids are truant or not. I can waste a lot time chasing after a few kids. I can't go out on the street and physically drag a kid to school so I don't have much power even though state law says the kids have to be here. Then I find out from the police, by accident in a casual conversation, that this one kid had been to court for some offense. I can't remember what. The judge put him on probation and made coming to school a requirement! Here I have been chasing after this kid for weeks. Do you think anybody informed the school that this kid was supposed to be here or checked to see if he was? No. Well, now the kid loses more respect for the law, and he's probably going to get in more trouble. Why would he care? No one's checking to see if he goes to school like the judge said so he's thinking, "They won't do anything if I get caught again." It doesn't make any sense, and it makes me mad!

When a teacher finds out that a child's family is struggling to put food on the table, a social service agency could help. But maybe the parents are immigrants who won't come into the school, and the teacher never mentions the family to the guidance counselor who knows the person in the welfare department to call. So the teacher feels bad, but nothing is done. The larger and more bureaucratic the school, the more likely the information doesn't get to the right person.

If all children are to be educated well, school and community must work more closely together. The case of one teenager illustrates what happens when, although professionals in each satellite are doing the best they can, they remain disconnected from each other.

> A depressed, pregnant, drug-using teenager in Maryland saw three counselors each week—a suicide prevention counselor, a parenting counselor, and a drug abuse counselor—and none of them talked to the others. All the while, the student missed so many classes that she flunked the semester and dropped out when her baby was born.[27]

Clearly, we need to find a way to forge links among parents, schools, and all the community agencies and resources so that children's needs can be addressed. Children will be the losers as long as parents, educators, and other professionals continue trying to address individual problems without the knowledge and assistance they could provide to each other.

Parent and Community Activism and Advocacy

Although educators would welcome activists working to gain support for what schools are doing, by maintaining relationships with parents and community members following the traditional paradigm, they inadvertently foster resistance. Scholar Michelle Fine, calling attention especially to the lack of power urban parents have, describes the current situation more graphically: "Parents are being promiscuously invited into the now deficit-ridden public sphere of public education, invited in 'as if' this were a power-neutral partnership."[28] She notes, too, that both progressives and conservatives are "distressed" by the failure of school reform. And perhaps surprisingly, they are both pressuring schools:

> Sometimes parents are being organized as advocates for their children, other times as teacher bashers, often as bureaucracy busters, more recently, as culture-carriers, increasingly as consumers. Parents enter the contested public sphere of public education typically with neither resources nor power. They are usually not welcomed by schools, to the critical and serious work of rethinking educational structures and practices, and they typically represent a small percent of local taxpayers.[29]

Yet as many schools learn, when people feel they have no voice, when their participation is limited or controlled, or when their values, goals, or family budgets are threatened, they respond. Parents protest new programs. Taxpayers vote to cut the school budget or to oppose the bond issue for a new building. Worse, when some unhappy parents or citizens begin talking with their friends and neighbors, they gain more power to resist by joining together. Two stories in chapters 3 and 4 show how organized opposition created serious problems for educators in two communities.

On the other hand, schools that take seriously the importance of building ongoing, two-way communication with parents and other community members seek out and listen to the differing ideas present in their community. They find that, because a foundation of trust and respect has been developed, parents and community members generally will be more prone to work with them to solve problems and to implement new programs. Chapters 7 and 8 present portraits of two schools that describe how the teachers, administrators, and parent leaders worked hard to create a climate of openness.

There are times, however, when parents and other community members play an important role as "watchdogs." When schools dismally fail the children they serve, community-based organizations can—and probably should—organize and put pressure on school boards to bring about needed change. Unfortunately, many times their efforts lead to ugly confrontations between the community and the school, generating a circle-the-wagons mentality among educators. In chapter 5 we present a different way of thinking about the home–school–community relationship, and in chapters 7, 8, and 9 we provide examples of how parents, community members, and educators can join forces to reform schools as *allies*, not adversaries.

The old paradigm where home, school, and community exist as separate, independent satellites no longer works. Schools must do more than help children develop academic skills and knowledge. The bureaucratic organizational structure with hierarchical relationships is undemocratic. Families and students are not treated equally; cultural and social differences are either ignored or erased. Only some parents feel confident enough to attempt to negotiate on school terrain. Educators, too focused on preserving their power, view parents as problems. Defensive and self-protective, they control and limit the access outsiders have. Yet when they need parent or community support, educators do not hesitate to ask "What can you do for us?"

Maybe it doesn't take a whole village to educate a *privileged* child, but it's easy to see that using the resources and riches in the entire community would make education more efficient and, most important, more effective. To ensure that *all* children will learn cannot be the responsibility of educators alone. When we reconceptualize the home-school-community relationship, we will ask: "What can all of us together do to educate our children?"

That would be the ideal, but the reality is that sometimes educators don't even think much about parents or community members until it's too late. That seems to be what happened in the two communities whose stories appear in the next two chapters. Yet, as these accounts show, a crisis, although painful for everyone involved, can lead to some positive results.

3

FROM CONFRONTATION TO CONVERSATION: THE STORY OF THE SOUTH WELLINGTON PARENTS' SPECIAL-EDUCATION GROUP[1]

The spring of 1998 was one of conflict and tension in South Wellington, a small New England town with no history of violence in school. Yet two months after the school shooting at an Arkansas middle school, South Wellington experienced several incidents that worried local police, school officials, and community members. At the same time several parents of special-education students were upset because the school district was not providing the resources their children needed. In the winter of 1999 these seemingly unrelated concerns escalated, converged, and were openly aired at a very emotional and painful public meeting of unhappy parents with cautious and somewhat defensive school administrators.

Understanding the connection between school administrators' concerns about school safety and parents' concerns about how their children were being educated and disciplined helps explain why parents and school administrators had such contradictory perspectives on what happened. Although it is true that threats against a principal and school vandalism are not on the same scale as a student bringing a loaded gun to school and using it, as was the case in Arkansas, people were concerned. When action was taken against several students, reporters were there to get the story. And the story they got made people even more nervous—and some parents, very angry. For instance, in an effort to ensure the safety of all students, school officials worked with police as they attempted to solve problems with students. That meant reporting the names of suspended students to the police.

Some parents felt the school department was going too far, branding children as "criminals" over what they believed were only issues of misbehavior at school. They did not think what happened called for police intervention. Rather, the school should have been addressing these students' special needs. One parent, Diane Davis, who acted as a volunteer advocate for a 14-year-old student whom the police were planning to charge with terrorizing as a result of one of the junior-high incidents, was quoted in the local newspaper as saying "He's a great kid, a bright kid. He wants to do the right thing. He needs help."

In addition to helping this student as an advocate, Diane also was involved in a dispute with school administrators about the educational placement of her own son. Even though there had been a written agreement between the school district and the parents to pay for one year of private schooling for the student, after which he would return to a program in the local high school, his parents wanted the school district to pay for him to return to the private school for a second year. School administrators stuck by the original agreement because prior appeals of their decision had been decided in the school district's favor. The parents wouldn't take no for an answer. The case continued through the circuitous and complicated path of due process: The parents filed a suit against the school district in the spring of 1999, and the school followed by filing a suit against the parents.

Diane began to connect with other parents of special-education students who thought that their children were not being served well in local schools. Thus, many parents, not just one, began to question school district decisions. School administrators seemed to be under attack on several fronts.

How and why did the relationship between the parents of special-education students and educators become so tense? Were administrators as uncaring about children's education as parents believed? Or were there two reasonable sides to this conflict? And what happens after a conflict reaches the breaking point in so public a venue? The story of the conflict in South Wellington shows that confrontation can evolve into conversation. Over time people whose very different perspectives cause them to disagree can develop a more positive working relationship, but getting there is not easy.

Diane was perhaps like other parents of children with special needs, who often depend on educators to help them understand why their children have social, emotional, or academic difficulties and to provide school programs that will help them learn. Yet because learning is so complex a process, influenced by many factors ranging from a personality conflict with a particular teacher to dyslexia or uncontrollable hyperactivity, even putting a label on a problem, such as ADHD (attention-deficit-hyperactivity disorder) does not mean that there is an easy answer. Thanks to scientific research, more is known about the brain and how it functions today than once was the case, but no one can say with any certainty exactly what conditions or strategies will work for any particular child. Thus, educators and parents must often resort to a trial-and-error approach with each child.

Yet even when both the parents and educators are doing the very best they can, some children still will not succeed. And parents, who have a deep emotional investment in making sure their children get a good education, cannot always be patient when they think that what the school is doing is not helping. Parents know that a good education is the key to a child's future. So when children come home every afternoon in tears or beg every morning to stay home because of a headache or stomachache, they want action *now*, not tomorrow or next year.

Conflict between educators and parents over the placement of special-education students happens everywhere. Laws and regulations governing special education, in fact, spell out a detailed process through which disagreements between parents and educators can be resolved. This process, though, is hierarchical and becomes more and more adversarial at each level. Finally, if there is no agreement at any earlier point, parents end up, as Diane Davis did, confronting educators in court. By then, both parents and educators have invested a great deal of time and money without having done anything to help the child.

In addition to the expense of the legal wrangling in these situations, school administrators in South Wellington found they had other problems. When the unhappy parents began connecting with each other, they shared stories of how the school district had failed their children. Coffee-klatch conversations can become a feeding frenzy as parents play off each other's concerns. Yet when parents cite specific instances and ask school officials to respond at a public meeting, as they later did in South Wellington, officials really can't defend themselves. School administrators can say very little without violating the confidentiality of individual students. What becomes common knowledge from media reports and community gossip, then, is primarily the story as told by the parents, which may be a skewed version of reality and certainly not what school officials would say if they could talk. As the South Wellington special-education director explains, "We have always been painted by parents in the ways they perceive us. We can't defend ourselves because of confidentiality. We don't get a fair shake in the public eye. We would be hurting the child if we did."

Despite their competing and often contradictory perspectives, South Wellington parents and school administrators eventually were able to move beyond conflict and public confrontation to conversation and cooperation. While individual issues still exist, relationships have improved and there have been changes in the special-education program. Rather than waste money defending itself against lawsuits, school administrators have become more proactive, spending funds instead to help the children. Most parents, even those who were the most upset, now believe they can talk with school administrators about their concerns. The end result of a conflict can be positive. But because the process was so painful for everyone involved, the South Wellington story may be helpful to parents and school officials elsewhere. By understanding more about these differing perspectives in South Wellington, readers may see how to gain needed change with much less pain in their own communities.

SPECIAL EDUCATION RESOURCE GROUP: THE BEGINNING

The public meetings in the winter of 1999 were sponsored by a parents' group that now has a name, the Special Education Resource Group, and meets regularly throughout the school year. At first, though, the group was informal and had no name. It began when several parents sat around the kitchen table in the home of Diane Davis in the summer of 1998.

Diane, who was struggling to get what she thought were necessary services for her son, began to hear about other parents whose children also did not have what they felt were appropriate educational programs in South Wellington schools.

> I . . . was trained as a parent advocate for school issues. And I was continually getting calls from parents . . . in this area. And then I kind of made some contacts . . . with different people in South Wellington, and we found out that we had common threads. . . . [M]y sitting in the principal's office, . . . being told . . . you know, "Mrs., you're the only one having these issues," was kind of isolating and making me frustrated. But now I was coming to meet other people [who had the same experiences].

The word spread, and Diane and other parents began to share their stories around her kitchen table. At some point, the group decided that, since there were so many parents with similar concerns, they should band together and do something to bring about needed changes. Their children not only deserved no less, but also each child was guaranteed by law an appropriate public education. The small group organized itself as the Special Education Resource Group.

The organizers began contacting other parents they heard about and put announcements in the local paper and on supermarket bulletin boards. The meetings of interested parents were held on neutral ground in the conference room of a local bank. Ironically, the parents say with a bit of embarrassment, the meeting room was not handicapped accessible, but, fortunately, no one who needed such accommodations came. At first, the meetings, intended to provide support and information to parents of children with special needs, attracted only about a dozen people. But the numbers grew. Finally the group decided in the winter of 1999 to invite school officials and school board members to a public forum. Parents hoped to get responses to a list of concerns they had compiled and provided to the superintendent and the special-education director ahead of time.

About 35 to 40 people showed up at the first public forum on February 11, 1999. An administrator says, "[When] we walked into the meeting, we were walking into a den of angry lions." The stories told by some of the parents show why parents were so upset. As readers will see, everything seems to be the fault of the school. As the special-education director commented when she read the draft of this chapter, "They [the parents] don't give the whole picture. The school did many things the parents don't mention." Therein lies the central problem in South Wellington: Parents and administrators saw things from very different perspectives. And, when there was no open communication, the problems got worse.

The parents' views follow; then the administrators share their perspectives.

THE SITUATION ACCORDING TO PARENTS

As noted, Diane Davis, the leader of the parent group, went the full route of due process, including later filing a formal lawsuit against the school system; but several others, who hadn't hired lawyers, also tell painful stories about their repeated—and unsuccessful—attempts to get the school district to provide what they believed were essential services for their children.

One reason parents cite for the failure of the schools to address their children's needs was that administrators treated students' behaviors as discipline problems rather than as indications that their academic placements were not working. Thus, students ended up being punished for what they did. No one seemed to consider the inappropriate behaviors as symptoms that should have been addressed through the special-education process. As one parent says, "They set them up to fail. They did not give them programs, individual programs, [which] they were supposed to do under the PET [Pupil Evaluation Team], the IEP [Individualized Educational Plan]. What they were supposed to do, they didn't! Then when the kids started acting out because they didn't get the proper learning tools, it was just this chain reaction of constant disciplining and sending them home instead of trying to help them."

Diane recalls what happened to JR, her son:

> I had been a very active parent since my child came to South Wellington in the 3rd grade and kind of handed him over to the process. . . . I thought that they were doing their job and that he was okay—seemed to be okay. [But] when it came to middle school . . . , he started to exhibit behaviors and I then started doing my homework. I took a parent-advocating course to try to figure out what was going on in the process. . . . (My background actually is special ed. I'm an ed. tech 3.) When I had an evaluation done, we discovered that with intensive special ed, this kid had managed to only go one year in his reading! And he was now facing sixth grade reading at maybe a second-grade reading level. And the behaviors were a result of him not being able to function in class. And not being able to meet the demands the teachers were putting upon him. He was avoiding and acting out. [Many of the calls I get] from [other] parents have middle-school students because they're not ready [for that level of school]. Schools had not gotten them ready to be in that place, to be independent. And they're not reading at appropriate enough levels to be able to be successful people. . . . They [the children] were astute enough to recognize that they're not successful and they can't measure up to their peers. And that's when their behavior started.

Another parent points to Brenda Sadler's son as another example of the way the school mishandled problems. Brenda notes that Larry, who has Tourette's syndrome, first started "acting out" in eighth grade, but the problem got worse in high school. Yet, as Brenda explains, the school did not recognize that a medical condition might be responsible for his behaviors until years later:

> We didn't have trouble until the seventh grade when my son told me, "I can't read books anymore. I don't remember what I've read." So when I went to the teacher and told her he can't read books, she tested him [and the

test results showed that he was reading above grade level].... She said, "Oh, he's pulling the wool over your eyes." So he went into the eighth grade, and I guess that's when ... when Tourette's, going through puberty, things start changing, the work is changing, ... your study habits have to change. And Larry had a high IQ so they would always say, "Well, he's just *choosing*." That was the big word. Larry is choosing to do the work—or not to do the work.... [But] Larry's psychologist would say, "You know it's like telling somebody who's crippled, just get up and walk a little, and they would!" Larry had a neurological disorder, [but] they [thought] ... because he had one good day, they expected that everyday he could do that. But that's not the way it is.

[It was only] about a month into high school when I found out he wasn't even going to class, he hadn't done any work, he was flunking everything, and they were calling me and suspending him. He wasn't in class—he was roaming the hall. He never did anything really bad.... It got to the point when they were sending him home all the time on suspension, and I said this kid has to be learning something. And they said, "Well, Larry doesn't want to learn." [But I thought] it's because you haven't created an environment where he can learn, because he couldn't sit still. He had to get up and walk around and relieve his tics. It came to the point that I think they just wanted to ... to send Larry somewhere, and I wanted him to be able to go to his own high school and have his own education. So I was actually having the opposite fight that everybody else was. They *wanted* the school to send their kids somewhere else, and I wanted him to have an education there. So it really became really adversarial, and that's the whole key.... [I]t was this adversarial environment.... There were two sides: It was "us" and "them." They were blaming parents. We were beside ourselves. We were looking at the school for help, and they were looking at us to tell them what we wanted for our child.

And Larry also had oppositionality disorder. So there are certain ways that you would approach a child who has oppositionality ... and he had some compulsive disorder.... They just did everything wrong for a kid who had these disabilities. [But] they didn't have the tools or the people there that had the right expertise to teach these kids. I said, "Why don't you have somebody who knows about learning disabilities design programs for these kids?"

I'd go to a PET and I'd say, "What would you like in Larry's IEP?" ... They would say, "What would you like to see ... ?" My husband and I would look at each other. We would try to brainstorm what we thought he would need, like you know, shorter increments of time, but we didn't know what we were doing! Then senior year they finally did hire a psychiatrist who has a daughter with Tourette's syndrome.... That's [finally] when things started getting a lot better. They really seemed to realize that this kid has real neurological problems. It's not just that he's acting out....

The parents viewed the message from the school as being "We can't educate the child now because the child is acting out." Instead of figuring out what children needed to learn, the school was "blaming the victim." JR, for example, was supposed to get help, but, as his mother, Diane, explains, that didn't happen the way it should have, and she blamed the school:

He had modifications that said that he was supposed to be given copies of class notes because he has severe learning disabilities and processing issues. He couldn't hear, listen, and write, and get it all down at the same time. So I kept calling and saying to this one teacher to please get him copies, and she wouldn't return my phone calls. And what would happen to my son instead was he'd be in a group and his teacher would say, "Where are your notes, JR?" And instead of JR's saying "My IEP says you're supposed to give me notes and you haven't," he would say, "I don't know." And the teacher would keep pushing his button, and eventually he'd say "Screw you" or something, and he would be out in the hall. And this would be a continuing pattern and I'd go to the principal, thinking there'd be some reasonable explanation for why they weren't following his IEP, and he would say something to his teacher. It became this ongoing battle of what is this IEP? It's useless.

According to other parents, JR wasn't the only one who wasn't getting the accommodations described as necessary in students' IEPs. Their perspectives were that the regular-education teachers too often didn't know what modifications were required by children's IEPs, ignored them, or refused to make them. In JR's case, however, the situation escalated.

He had this test one day, and he was supposed to get his own test, but nobody else was supposed to know he was getting a different test.... [T]his is what happens, too: We identify these kids in front of their peers, and they isolate him immediately. The minute the cat's out of the bag, because we haven't educated our kids that it's okay to have disabilities, so he's sitting there with no test in front of him, and the teacher said, "Oh, I forgot to give you your *special* test." So he gives JR his test, and one of the accommodations JR is that he is supposed to have somebody read the test to him.... But the teacher doesn't read the test to him. So he keeps going up every time he gets stuck, and the teacher one time told him to go sit down. He couldn't help him any more [because] he was busy with somebody else. So my son stood there and at one point [the teacher] just got frustrated with helping my son and threw him into the blackboard. All the way across his back, a nice big bruise.

Diane reported the incident to the police and followed up with the school. In the end the teacher, who was a substitute, was let go. When, as Diane had requested, the teacher apologized to JR, he said, "I just never knew how to help you all year." Even though Diane was somewhat sympathetic, that didn't change her outrage at what had happened to her son.

> [He understood] that it wasn't that this kid was acting out; he had severe learning disabilities. He couldn't read the material. Even some of the stuff he could read, he couldn't process . . . couldn't understand it. And this was this man's frustration . . . the same frustration the parents face. But we don't come in and take a teacher and heave them up against the wall. So I think the school realized that they were in a bad situation. JR had made no progress, which is called regression—he'd gone backward. And then this happened. . . . [T]hey agreed to place him in an [outside] placement for a year. A special-purpose school. And he made enormous emotional gains—you can't even measure those. . . . Academically he made a year's gain in his reading. He hadn't done that in five years' time. He was just feeling good about himself.
>
> . . . [But people from the school district] never once called, never once came to view the program [in the private school] or even find out how he was doing. Never once. But I was happy because I knew he was progressing, feeling better about [his] being a student again, and there was hope that he would learn to read. And so they [South Wellington educators] came to his end-of-year meeting . . . [without] a copy of his IEP . . . and said, "We've designed a program for him up at the high school that's perfect for him. We're taking him back." And it's like, well, we've just stepped over the process, because the process is—we first have to look at the student's needs and talk about how to meet those needs and the *last* thing we talk about is *where* the placement is going to happen. But they came, storm-trooped in, said, "We're taking him back, thank you very much, good-bye." And everybody was shocked. And I challenged that because I wanted him to at least stay [at the private school] a half a day the next year. . . . [H]e was just starting to feel good about himself. He was just starting to make academic gains. Why are we going to yank him out? I wanted him in his public school a half a day, too, but they wouldn't agree to that so that's where all the power

struggle happened. And it went to what they call due process, which is very, very emotional for my whole family.

The next fall JR went back to the local high school, but all did not go smoothly because, his mother says, JR did not get the modifications he needed—and then there were threats about calling the police:

[I]f he got frustrated sometimes, I mean academically, he would just need to take a step back, take a breather, take five minutes. These kids just aren't often as resilient as other students. And so the plan was that he would kind of remove himself to the back of the room . . . [when he was] getting frustrated, he couldn't take any more "process" . . . , or the teacher would give him a visual sign. . . . [T]hat was the plan we had all agreed on, and it was working. [But] then the teacher decided—for whatever reason on her own—that she didn't like the other students seeing that. But . . . that was his individual plan. He's in a self-contained room; this should be a safe environment for being able to do that. [Even though] she wasn't allowing him to do that any more, he kept doing it because he knew that was his plan. That was what worked for him to bring himself back. Well, eventually she kept calling in the behavior teacher from across the hall, which became a very threatening activity for JR because he felt he was one more step from the behavior room itself, and that's the end-all: You know you're here, and then you're in the behavior room. Boy, you're gone! And so this [other] teacher came in one day when JR was refusing to do his work. He said to him, "If you don't do your work, we are calling the police." The police! This was never denied, and for a kid like JR, this was real for him, knowing everything else that was going on.

Then he was in a health class one day and part of his behavior plan is that, if he's working for 20 minutes, he can take a self-imposed break. So he had raised his hand several times, and the teacher wouldn't let him go. It wasn't that he wasn't doing his work—he just wasn't going to let him go. And JR very boldly in front of his peers, you know this was a regular ed. class, said, "I know what my IEP says, and I need to take my walk now." He went down the stairs to the boys' room [bathroom] and by the time he almost got . . . [there], six teachers were chasing him with walkie-talkies. And all this talk about police and all this other stuff that was going on. . . . They pursued him throughout the whole building, and then he ran out the door. He ran all the way to Annette's [another member of the parent group] house. He was a mess.

Diane goes on to explain the "bad communication" between her and the school at the meeting held to discuss her son's situation.

When my son had run out of school, we had that emergency meeting. I'm sitting at this meeting . . . ready to fall apart, and the assistant principal slid over this sealed envelope to me in the middle of what I was saying. I didn't look at it, but when I got home, it was my son's suspension notice. [At that point] I just withdrew him from school. . . . The only person who it [the letter] was c.c.'ed to was an officer at the police station!

At this point JR's parents went to due process. Diane was teaching him at home while he was "in between placements" when one day a sheriff's car pulled up in the driveway.

. . . [W]e had already filed due process, just kind of waiting for the whole thing to happen, and what have you [when] the sheriff served me papers for due process from the school! Now, I know enough people up in the [state] department of education that I felt a phone call was appropriate, so I called the due process coordinator. I said, "Mark, I know this is probably inappropriate, but I'm gonna ask off the record, what's going on here? I just got served due process papers from the school during a due process." And he said, "Diane, it's called retaliation. I've never seen it [rise] to this level ever before." My son literally cried. "[A]re you going to be arrested because you're going after my rights?" I said, "No, that will never happen." Even though it [the suit] got dismissed, it was intimidation.

Other parents in the small core group describe situations with their children that reveal some of the same concerns that JR's and Larry's cases illustrate. In each instance, children did not get the services they needed. Parents began to lose faith in the school system, its administrators and teachers. Too often, they say, teachers didn't even know about the children's IEPs and didn't return their phone calls. Administrators and teachers used discipline policies rather than the special-education process when the kids "misbehaved." Some even began to get paranoid. One member of the group used the phrase "crazy-making" to describe actions taken by school officials in addressing the parents' concerns. After meetings about the children, parents felt alone and isolated because they were made to believe that, as one parent says, "We all thought we were the only ones with dysfunctional kids in the school system." Another added, "[These were] the worst-acting kids. . . . [E]verybody had been told that nobody else had the complaints they did. And then . . . [these parents] discovered each other." But, even worse, as one parent explains, was that the school would not admit there were problems: "My son was having difficulties, and they kept saying, everything is fine at school, we don't see any problem. [When] the guidance counselor [called], she said, 'What's going on at home? Because, think about it. There's nothing happening at school. Everything's fine. So what's going on at home?' [But there weren't any changes, no divorce.] . . . We're happily married. We're Ozzie and Harriet Nelson, for God's sake."

This group of parents educated themselves about the special education process because they believed they had to in order to get needed services for their children. One parent remembers:

> I didn't even know what the rules were. I didn't know that I could go above the teacher to the principal. . . . Don't you think that somebody's who's already on the team who has that information should share that with a parent? Don't you think? So when a parent starts to find out all this stuff on their own, separate from the team that already knows this information, don't you think they'd feel a little, well, why didn't they ever tell me this stuff? I mean it creates that us-and-them thing, and it's frustrating. And, on top of that, if a parent finds this child is now in whatever grade reading, you know, five years behind, that just adds to the fact that you've never involved me in this information before. That makes for some anger, some distrust. I want my kid out of here. This is damaging. . . . [P]arents become very protective at that point and say, "My God, I didn't even know this very basic thing!"

Another problem, the parents say, is that some parents of children with special needs have disabilities themselves, a fact that school officials do not recognize.

> I guess speaking for myself that a lot parents who have children with disabilities have disabilities themselves, learning disabilities. It's very, very hard to sit in front of the faculty who are all college-educated people when you may have a disability yourself and struggle with some of the same things your child struggles with and you have to fight his battles. And I think that that is probably more common than not, but people don't want to admit that. And it's very intimidating, and a lot of times you really don't have the knowledge to know how to fight this. They're much more savvy than you. They know the ins and the outs so, unless you have the financial capability to hire a lawyer to do the work for you, or a consultant, you really are like a lamb going to the slaughter.

As the parents talked about their experiences with their children, most thought things had been okay in elementary school. The real problems seemed to begin when the children reached middle school. It's easy to see why that is the case.

First, adolescents not only have special needs, such as learning disabilities or neurological disorders, but, with the onset of puberty, they also have to deal with raging hormones and a desire to become more independent, to rebel against authority, and to win the approval of their peers.

Second, instead of one or two teachers who know them—and their families—they may have seven or eight adults to deal with in the course of the day. Maybe this also accounts for what one parent called a change in attitude: "[T]he attitudinal change from elementary into junior high and then high school is really noticeable.... [Y]ou never feel like you intruded as an elementary parent. For the most part, you don't in junior high [although] it depends on the group of teachers or who's there, but you can feel like you're being really intrusive when you go into the high school."

Communication with teachers also becomes more problematic in the secondary grades. Even though she was supposed to get weekly progress reports, one parent complained that she didn't get them until she called the school:

> [My son's] so far behind [because] they go from the teacher to the guidance to whoever. By the time I get them, they're actually dated three to four weeks prior.... [P]arents [instead of someone at the school] often [have to] become the case managers.... I actually had a meeting with my son's case manager there, and she's very accommodating, [but] I said to her, "... you know I have ADD. I can't be baby-sitting his program. I am the last person that should be baby-sitting. I can come to these meetings and tell you what happened, but I can't be the one to implement it." ... [And] this is the silly part! She was his case manager, and she wasn't getting any of the progress reports! I was getting them at home. What am I going to do? His case manager's supposed to be working on his organization and study skills, which is his deficit, and his reading fluency, and where he is in his program, and she wasn't getting any of this [information].

Finally, the expectations for academic work are much greater at the secondary level. Adding to the growing frustration when students can't do the work, they also worry about being called "stupid" or "retard" by other teens. Neither teachers nor parents should be surprised when the kids act out. As one adolescent male told his mom, "I'd rather have her [the teacher] think I'm naughty than to think I'm stupid." Another parent says, "They'd rather fail their regular class than be in special ed."

These parents share a goal with which school officials would agree: to make sure all children learn and do well in school. Since these children had been identified as having special needs, their programs—including the services and resources needed to help them learn—were specified in the IEPs. Yet, for a variety of reasons, the programs planned by the team (PET) often didn't match what actually happened. Getting things down on paper was only the first step. One parent says, "You have to dog these programs. You have to keep calling and saying, okay this isn't happening." But parents don't want to be put in that position: "We have better things to do with our lives, and we want our kids to be independent learners. But we're forced to, and then we become the busybody parent because we're always in there." Another parent adds, "And that's a catch 22, because, the more you call, the more they think you're pampering your child." And parents worry that being too involved will make things worse for the children. As one says:

> And sometimes there's retribution for that. If you come in and you say something about something that happened . . . somebody [would] bend over to [my son] and say, "I've heard you went home and told your mom this." I mean, my son should freely to be able to tell me anything that goes on in his day. If he can't, then there's something wrong in that school. [But] he started to develop this issue about . . . maybe I shouldn't tell Mom when my teacher's really nasty to me. And that's not healthy because my son is not in a safe environment at that point.

When the Special Education Resource Group first met in the fall of 1998, parents' stories and concerns were the focus, "the pouring-out kind of stuff." But the group soon moved beyond the venting to organizing and communicating with other parents—and educating themselves. Parents had found out that they needed more information so they brought in outside speakers. Stan Cooper thought this aspect was "awesome." Grace D'Amato says, "[T]hat's one of the things I found so attractive about the group is that you go, you tell your story—we've all got terrible stories to tell—[but] you've got to get beyond that and do something positive and educate ourselves. That was the focus of this group and that still is." Sadly, what Grace gained from participating in the group didn't help as much she would have liked because her son ended up dropping out of high school.

The first year the group made time for people to tell their stories because "people really needed their chance to vent because they couldn't say anything to the schools. They were being really stifled." Hearing the individual problems they all had, the parents decided that they would like to have a conversation with school officials. Thus, the stage was set for two open forum meetings with administrators and school board members.

THE PUBLIC FORUMS: PARENTS' PERSPECTIVES

The Special Education Resource Group members carefully planned the first public forum. One parent explains their thinking, "We're going to be very fair about it. We were going to forward the special ed director and the superintendent our questions and concerns." Two members of the group would share the responsibility for facilitating the discussion. The group sent a three-page memo to school administrators. It listed specific concerns in five areas: communication; accountability; attitude; behavior management; and crisis prevention, intervention, and "at-risk" children.

However, the first forum didn't go the way parents had hoped it would. The superintendent brought, as one parent describes them, "all these men in trench coats—well, the principals, the police chief, the *police chief!*" [The superintendent later explained that the police chief did not attend this meeting. Rather, it was the commander responsible for writing a grant that would put a resource officer in the schools.] The plan was for everyone to sit in a circle, but "they [the principals and the police chief] sat outside the circle [behind the superintendent and head of the teachers' association]. I noticed that first off. They never said a word."

The parents hoped their "very specific questions" would be answered by school officials, but, says one parent, Superintendent Joe Ammons "starts with this spiel with the police chief selling the resource officer [a police officer who would be in the schools all day], and we're all looking at each other. 'Did we ask them to come do this?' It was awful. He was gaining support of his resource officer [instead of answering our questions].... The superintendent took control of the whole meeting and talked about everything except what we asked him to come for."

So at the end of the meeting the parents were frustrated. Even though the special-education director, Patricia Crandall, was there, "she didn't say a word." Diane Davis recalls, "[L]ooking through this list of questions, we've only answered one.... There were numerous people who didn't have a chance to express themselves. And ... so we invited him [the superintendent] to come back and finish the conversation."

For the second public forum, the group asked someone else to facilitate the meeting, got volunteers to take notes to see how many of their questions got answered, and also invited the local press "because we thought it would be kind of a good thing." The second get-together was much more productive.

This time, according to group member Annette Fisher, the special-education director Patricia Crandall "did a lot of answering.... [S]he clearly had left that first meeting very ... upset and hurt probably because she's pretty sensitive personally. [Then] she had gone and done some work out in the schools. She'd done some homework. She'd taken those questions [the list compiled by the parents] to the staff.... She asked them things and told them some of our stuff. She came back with some answers ... things she said she was going to propose the following year."

This time parents did have "a chance to talk," but "some of what they had to say wasn't fun to hear." The parents there mostly had issues, such as the ones described earlier, and they were emotional. They were, after all, talking about the ways in which they believed the schools had failed or mistreated their children. Another example that parents recall created some tension was the discussion about the policy of notifying the police when students were suspended from school. According to parents, when they challenged the right of the school to do that, the school district discontinued the notification.

Diane Davis, though, believes the result of the discussion in the end was positive: "Interestingly enough, ... all this upheaval, ... has to happen, you know, to get things to the surface." Annette Fisher, a retired principal who functions as the group's advisor, agrees: "[This meeting] was a turning point. They [school officials] recognized things after that." Diane heard "through the grapevine" that "word was spread among the schools from the superintendent's office that this group wasn't going away."

Annette adds, "Well, that's what I kept saying to him [the superintendent] every time I met with him.... Try to hold back all you want to, but this is a group of people who are not going to go away.... I got so mad at him. 'You're accustomed to stonewalling parents until they just throw their hands up and quit. I'm telling you this group is not going away!'"

School officials, however, were never as unsympathetic to the concerns discussed as many parents probably thought they were. There is even a reasonable explanation for what parents perceived as the superintendent's refusal to respond to their questions at the first public forum. The superintendent, for instance, later said that he responded to all of the concerns parents put on the list. The differences in the perspectives of school officials and parents may stem, in part, from their dissimilar roles and responsibilities.

SOUTH WELLINGTON ADMINISTRATORS' PERSPECTIVES

The school shootings in Arkansas and the serious problems soon after in South Wellington had administrators worried about school safety. The special-education issues seemed to get entangled in the behavior problems exhibited by a small number of students that worried school officials and local police. The special-education director says:

> The parent group got themselves going. It was based on negativity at first. They were not happy.... We had a group of kids at the time and things were not going well, not special-ed necessarily.... It was like a wave [of things happening at the same time]. People thought the school was not giving their kids a fair shake, but we cannot tell the public because of confidentiality. We walked into the meeting, and we were walking into a den of angry lions.... We didn't know what we were going to hear. What we heard was parents telling how the school had handled their cases. We could not respond very well because it was all confidential.... A lot of people there had obviously met before.

The superintendent describes his concerns about responding to the parents' issues in this public forum because of the people he saw in attendance.

> I had had to go to open court to get a restraining order against a student who had threatened another administrator with a baseball bat, and the police arrested him and had to use Mace. And I looked out, and one of those families was there.... Another parent of a kid who eventually was accused of setting the school on fire [was there]. These were pretty serious situations with students that we would deal with individually.... It was so unusual because we had several of them in a couple of months... not all at junior high.... I can understand why they [the parents] were upset. I would be upset if I had a 13- or 14-year-old who was arrested, and I had to go to the police station.

The superintendent may have been thinking of the eighth grader the police planned to charge with terrorizing, the student for whom Diane Davis was acting as a volunteer advocate. This student, as Diane Davis believed, probably did need special help. But school officials and the police then were focused on the safety of all students as they tried to address the escalating tensions and threats toward local teachers in the weeks following the high-profile shootings in Arkansas.

Because school officials had been working so hard for many months to deal with these serious problems and to figure out what steps might be taken to ensure the safety of students in local schools, it made sense for the superintendent to invite the police commander to this meeting to discuss the grant-funded plan for putting a resource officer in the schools. But because of his position, he also had another concern. As he looked around at those in attendance at the first forum, he knew he had to be careful what he said because he had a responsibility to maintain the confidentiality of all those involved in individual cases.

> Could the school do better? We could always do better.... [The parents there represented students who had had very serious problems in schools.] One parent we were in the middle of a lawsuit with. I was invited to speak. What could I say?
>
> ... Some were motivated to be there because they didn't like the good relationship we had with the police department.... I was bothered by [the idea of putting a police officer in the schools], but my attorney helped me with that.... [He made me realize that] we always have police at the high school for basketball games when there may be only a couple of hundred people, but during the day when there are 1,100 students, we don't think a police officer is necessary [then]? ... All of that was happening at the same time.... There were kids in trouble with the law inside and outside of school. That probably added fuel to the fire.... And some parents were very negative when we brought the chief and youth aide officer.... But with that group of kids, the media coverage, and what was happening all over the country then, we had to do something. We react fast.... Now that there has been another school shooting in California [another school shooting in 2001 in the San Diego area], maybe parents now don't think we are overreacting. We take immediate action. We have all those kids evaluated.

Both Joe Ammons and special-education director Patricia Crandall say that their goal at both meetings was to listen to parents. They would find things they could address and do something about. Many of the issues they mention are the same ones parents bring up, but they have a different take on them.

"Communication is important," Crandall notes. "Parents need to feel that we are really listening to them. It isn't just some cursory thing—we really are listening." She points out that the question "What are your [parents'] concerns?" appears right on the IEP form as part of the new special-education regulations.

School administrators realize that parents need information because special-education laws are always changing. One way the school district provides up-to-date information is by publishing a comprehensive special-education handbook, now in its second edition. Copies of the handbook are given to parents when new referrals are made, but parents also get a copy of a parent's rights statement whenever they get a letter from the school. But even with all this information, says Joe Ammons, "The problem is that the school interprets the law, and then the parents want more services."

When one parent of a high-school student said the teacher didn't know the student, Joe Ammons did some research.

> What was wrong with that teacher? Well, the student can make changes through add and drop and end up in a different class from the one everyone thought he would be in. The student ended up with a teacher who had just come in [as a late hire]. When you think about the way things happen in a high school, you realize that a student could be in a class where the teacher didn't know him. The parents, though, didn't want to hear any excuses. So we have been working to make sure that we follow up. There's a department head in special ed who goes and deals with those teachers where the student finally lands.

Communication with parents of secondary students can be difficult as well. Crandall says, "They can get progress reports fairly frequently, but it is hard at the high school for every week with the flip-flop schedule." Because with the block scheduling a class meets only every other day, a student could be in a particular class only one day in a week if there were an assembly or other change to the schedule. Joe Ammons adds:

> We have parents that want daily communication! . . . But maybe the teacher's mother died so there's a sub in the class. Then the parent says, "I didn't get a progress report from English." It's really hard to get parents to understand that. . . . And then there's the worry about confidentiality. We need to communicate enough so the parent knows what's going on, and it's reasonable for the teacher to do [when they have many students]. Parents

want to know any time there is a change. Special-ed parents are getting more knowledgeable, and I think that's good, but, at the same time, they don't understand what a school day is like.

Some of the parents think e-mail is an excellent way for teachers and parents to communicate, but concerns about confidentiality and protecting the privacy of students make the superintendent reluctant to encourage its use: "There is no confidentiality [with computers]. We try to protect the students. . . . We have a computer user agreement to do that. I just want to be careful that we protect a student's privacy. There is a phone in every room at the high school." Joe Ammons encourages teachers to contact parents by using means other than e-mail.

Ammons believes the special-education process itself sometimes makes communication—and reaching agreement with parents about appropriate services—difficult. Getting lawyers involved helps no one, and confidentiality concerns can prevent school officials from being open with parents.

> We don't gain much when it goes to court. They've got to pay their attorney. We haven't had so much of this in South Wellington lately. But it is really hard to have a good relationship with some parents. . . . If the parents are not happy, they can bring in an expert. Once that was someone I checked out and refused to hire in two previous districts. Now that person is coming in as the expert and telling us what the student needs. Parents trust the person they bring, but I can't tell them what I know. It's the same thing with outside testing. The way the law is set up now, we do our testing, but parents can pay $1,500 and have child tested. Then they come to us and say, "Now here's the bill." We can say no, we won't pay it, but then it will go to a hearing. If we say no, it automatically goes to hearing, and we have to pay our attorney. . . . That's the federal law. That's not set up to encourage teamwork. . . . The only people who make money are the attorneys. We've worked hard to cut our costs—and we've made progress. Let's work as a team and put the money to use for the students. I feel bad for the family. I don't feel a sense of accomplishment [even when the legal decision is in the school district's favor]. It's taken a lot of time. We didn't get into this business to do that.

The difference in the perspectives of school officials and parents will always be there. Parents' focus is always on one child; educators have to consider all children and the bigger picture that includes school district resources. And either can be blind to what the other sees as the superintendent explains: "People cannot see the reality of their own child. If it's a [behavior problem] or other things the child does in school . . . are not productive, manipulati[ve], they are blind to it. They care about their child, but they really don't see it." Understandably, parents may not want to be totally open about what happens at home—just as school officials can or will not share everything they know. Ammons says we have to get past the idea of blame: "How do we work together? That's been our major theme."

The forming of the Special Education Resource Group and the tense, emotional public forums have led to better relationships between parents and educators and, even more important, new services for students with special needs. Both the parents and school administrators point to several positive outcomes in the last few years.

SOME POSITIVE RESULTS IN SOUTH WELLINGTON

What began as confrontation has, over time, shifted the relationships between the parents of children with special needs and school administrators and teachers. In many ways both are trying harder to work together to find more effective ways to educate the children. The school district has become more proactive. Recognizing that money spent to defend themselves when parents challenge a child's placement helps no one, administrators now spend more money up front on evaluating the students. They have created new programs within the district so that fewer students need outside placements, and they have been quicker to hire outside consultants when such expertise is lacking in their own personnel. They have also made other smaller, procedural changes to improve communication and collaboration between parents and educators.

Patricia Crandall comments on what the school district is doing differently:

> We have been able to hire a psychiatrist to evaluate a child and to be a liaison between the family's issues and the school's to work out the best program for the child. We have been proactive in this. . . . [W]e get the initial evaluation and consultation done—let her look at the child, the family structure, and then work with the school—we don't continue paying for counseling. That has made a huge difference. It's expensive, yes, but it shows the parent right up front that we are serious about trying to help their child. . . . That's something that we've done that not too many schools do.

> [W]e have used the family focus group [part of a program at an outside institution].... The parent has to go down there two or three times a week to learn how to better deal with their child. Then they have social workers who come and work with family in home. The child spends eight weeks away [in the private school] with weekends at home. [Then] there is follow-up afterward. We have used that program several times successfully—not all the time.... The state Bureau of Mental Health is setting up more in-home support.... We have people coming in to inform school staff about services available. Social workers in the home.... There are many new services that were not available before. And they are not coming from our budget, fortunately.

However, as the superintendent points out, these programs require that parents be involved. If they refuse, there isn't much the school can do.

Not only is the district spending money to send students to a private school in the beginning, but, Crandall says, "[T]hen we have developed programs to bring them back. Being able to spend some significant dollars up front has helped us with the parents and made a difference for the schools.... We're really not all there, but now everybody is working together with the kid. We have been in a proactive mode since this major uproar a few years ago ... and kids have benefited."

Parents were quick to notice a small but, to them, important change at the high school. As Beth Jackson explains, "[T]here were complaints that she [the special-education director] heard at that meeting they did implement, at least at the high-school level. Each special-ed student's IEP would be provided to the teachers before the workshop days at the beginning of the year. They were required to have reviewed it, and I think that was a big thing. Yup. And I mean that was a direct result of her [the special-education director] coming to the group and hearing the complaints." Grace D'Amato adds, "And there was even time set aside in the workshops before school started to go over these.... [I]t was made a priority." [When special-education director Crandall reviewed the draft of this chapter, she noted that teachers do not get the whole IEP, which would not be very useful for them. Instead, they get copies of the relevant parts of the IEP that will help them make the appropriate accommodations for their students.]

When the school district brings in experts to educate the staff about particular problems, interested parents are invited. Parents appreciate also being invited to special-education inservice programs. They also note that, when Crandall revised the parent handbook for special education, she sent a rough copy to the group. Parents had hoped the business card for the Special Education Resource Group would be included in the handbook as an insert, but, so far, that seems not to have happened. Thus, group members have been handing them out themselves at school meetings because they want to reach more parents. As Stan Coffin says, "That's why we've got to get organized because we're there, we're there with them. That's the big thing." Diane Davis says, "[W]e need to get that out there to all those parents. We need to let them know there's a parent handbook the school puts out. We need to let them know there's a support group. We need to let them know the school's working with the support group. You know that stuff has to be shared."

The special-education director seems to be helping the group. As Diane recalls, the parents were surprised—and very happy—when they heard Crandall at a meeting: "And Patricia was amazing at that meeting and the things that I think she said. . . . [W]e need to tell teachers that advocates at meetings are good things. That this group is a good thing. I mean, she really said some positive things about working with relationships and stuff. . . . [A]nd for her to say that word *advocate* and to say *good thing* in the same sentence is amazing because I know she was always very intimidated by having one there."

The parents do seem to understand how difficult all of this is for Crandall. Stan Coffin asks, "Can you imagine that woman's day?" Someone responds, "It's got to be hell." Stan thinks that advocating for the parent group actually can be helpful for her: "The ignorant parent about what's going on is her worst enemy because that person just has feelings. They [the parents] get hurt, they don't know what's going on, they get mad, and the whole thing isn't going the right way." The Special Education Resource Group can help educate parents about special education.

Crandall also has encouraged special-education staff members to attend meetings of the parent group, scheduled from 6:30 to 8 on a weeknight, "so they wouldn't run on and on." Annette Fisher explains that teachers didn't come to the first meetings of the group because "there'd never been a message that they should be involved with us until this year. . . . When Patricia changed her message, people began to come. "Another parent adds, "Yep, she said to her department heads, this group is a good thing, and we need to work together."

Crandall says she does want teachers to be involved, but it isn't always easy: "I've encouraged my staff to go to some of their evening meetings, but it is hard to have a balance. For instance, they've invited an attorney [to an upcoming meeting] to tell them what's in a good IEP. They should have an educator telling them that. This puts us on the defensive. [And because of what has happened in the past] the staff feels on the defensive. They are really fearful."

The parent group, too, has made some changes. Parents know that teachers work hard at a very difficult job for which they get too little recognition. Thus, the group set up and publicized a process for recognizing some of the teachers, as Annette explains, "[To be recognized], all it had to be was one parent saying that . . . [one teacher] made a difference for their kid and to write a thank-you letter. And it might mean that somebody would get a thank-you that would make the rest of us clench our teeth, but, if it made a difference for one kid, then we could accept that."

Nomination, or thank-you, letters—14 in all—were sent to Diane Davis, who was the only person in the group to know who sent them. Diane announced the recipients of the awards to the group, and members shared the responsibility for presenting teachers' awards at individual school faculty meetings; Annette gave the special-education director her award at a school board meeting. Each person recognized was given a framed certificate, "Teachers Making a Difference," created by group member Beth Jackson, and a bookmark with a poem. Since these recognition awards also got publicity in the local paper, the Special Education Resource Group gained some positive notice in the community. Members hoped that the general perception of the group—that they were trouble—would begin to change.

The group also realized that other steps would help everyone move forward in a positive way. Bringing everyone together to develop a vision for special education made sense because, as Diane says:

> We're not going to go anywhere as a team if we don't draw these teachers and these administrators somehow to make things better. We can sit here and gab all about what should happen, but, if we don't have the vested interest of these people, the other players, then it's not going to make a squat of difference. So we decided our theme this year is partnership and communication. . . . We started our initial meeting with the vision, the PET vision, what it currently was for everybody who attended meeting whether you be a teacher or a parent. . . . [About 30 people, a fairly even split of parents and educators, attended the meeting.] And it worked out good because we were able to break down in groups and talk about what we saw our current vision being and what we would like it to be. . . . [I]t was very helpful, but it just kind of dismissed itself after that because. . . . We invited the same people back to work on the action plan, but we only had four or five people come back. . . . So it's kind of hard to get anything done when you

> **Sidebar 3.1**
>
> ## Teacher Recognition Award Bookmark
>
> On one side the bookmark given to recipients of the recognition awards had the words: "Parent's Award of Excellence, Kindness, and Understanding." On the other, this poem was printed:
>
> THANKS FOR...
> making the difference
> long, long hours
> creating a sense of family
> being the keeper of dreams
> using good judgment
> forgiving
> that sensitive touch
> never giving up on anyone
> believing in miracles
> respecting each other
> taking responsibility for all students
> being brave
> smiling a lot
> showing enthusiasm even when you don't feel like it
> keeping your promises
> giving your best
> your wisdom and courage
> providing creative solutions
> avoiding the negative
> seeking out the good
> being there when students need you
> listening
> doing more than is expected
> remaining open and flexible
> being a friend
> sharing
> being a child's hero
> going the distance
> having a good sense of humor
> being a dream maker
> giving your heart
>
> —Anonymous

only have minimal support. But I think that the school's at the point now, though, that they see us as very supportive of them being able to get their job done and get it done easier and maybe better....

Special-education director Crandall does believe the group is helping—although she still worries a little:

Parents really need a support system.... This group is trying to do that, but I think sometimes they slip back into blaming the school.... [It isn't helpful if a small group is sitting around complaining about teachers. Come talk to me or directly to the school and the person involved.... I wish they had some person with some insight about school moderating these meetings.... We could try to do something instead of just stirring the pot.... [A problem can be] the language people use when they are talking to parents. Staff members have to be really careful to make sure that they are not using any terminology that could be insulting to parents. Parent advocates who come in and help parents with that—people who come in as helpers—that's really good. [I want] staff participating in meetings at night—they really want to help. I have tried to impress on the staff to let parents know they are listening, get back to them right away. Getting back right away makes a difference.... We have a long way to go.

The superintendent adds, "We are a better school system for that resource group. We are a better school system because of our relationship with them. And the things that they have shared with us have a common theme. And we have definitely tried to change that even though they may not even realize it or notice the things we've done. Making sure teachers understand and know what the accommodations are..." Joe Ammons would like to see more parents involved—even those without special-education students. "We've come a long way. These kids didn't used to be in school. We need to educate people.... [The important thing is] what can we do to help your child be more successful?"

Because going the route of due process to get there is not helpful, the superintendent and the special-education director both agree that taking a more proactive stance toward figuring out what kids need has been working. Ammons says, "We are never going to be a winner in a battle. Even going through months of court issues and having a judge say, 'The school department's right.' So what? What's good about this? We still have the family, we still have the child, and they feel worse than when we started.... We have to figure out how we can make parents know we care."

SOUTH WELLINGTON TODAY: HAPPY ENDING?

Although the relationship between South Wellington parents and educators is no longer a public conflict, this story has no fairy-tale ending. With any students who are having difficulties in school—but especially with students who have special needs—one can probably never say "And they all lived happily after." These problems are human and so complex, so complicated, that no one—not parent or educator—has the "right" answers. What works well for one child may be exactly the wrong approach for another. And usually no one knows why that is the case. Much more research is needed on the physiological and emotional factors that influence learning and not learning and on effective teaching practices. Lacking a useful knowledge base, parents and educators must struggle, experimenting with different modifications and services to see what might help a particular child learn. As they weigh alternatives and options suggested by this consultant or that expert, they must negotiate and compromise. Sadly, there is no other way in South Wellington or anywhere else in America.

One South Wellington parent says the situation is better, more positive—even as she so far has been unable to reach agreement with school officials about her son's current educational program. "I won't give up," she says. "This is too important for my son." The failure to reach consensus in this instance once again illustrates both the differing perspectives parents and educators bring to the table and the effect these have on making decisions about the allocation of limited resources. Because the parents focus on one student, "my child," and view his or her education as *the* most important priority, they care less about the cost of a program to the school district than about getting what they believe are essential services to ensure that their child will be successful in school.

Educators, on the other hand, can't help thinking about the needs of one child in the context of what is best for many students, "*all* children." Thus, a teacher might worry if a suggested modification for one student might create problems with other students in the class who think he or she is getting "special privileges." Because even experienced teachers worry about losing control of a class, they are reluctant to do anything for one child that they aren't willing to do for everyone. The financial implications of providing services can be another sticky issue. The superintendent, charged with managing the district budget to serve all the students, cannot help but think about the costs of a particular placement. For instance, if the school district must pay $50,000 for one student to attend a private school, those funds can't be spent anywhere else. Even though the school district is legally required to provide an appropriate educational program for a student with special needs, it is not obligated to go with the most expensive option. So as the superintendent tries to balance the needs of the whole school system, there may be no way to avoid disagreements with individual parents. What either administrators or parents view as "appropriate," the other side can just as easily think is inappropriate or not enough.

South Wellington administrators did listen to parents' concerns, and they made changes—though not enough to satisfy every parent in every situation, an impossible task. The South Wellington story, which began in conflict, has resulted in a more collaborative effort to help students learn. And although the relationship is still fragile, both parents and educators are trying to work together. Communication is better, although not perfect. Both groups have learned how important open communication is to building a trusting relationship, and they seem committed to maintaining the gains they have made. The Special Education Resource Group formed to address the problems a few parents saw has helped many other parents. It continues to provide much-needed support, information, and education for parents who would otherwise be left to cope with difficult issues on their own.

So, in the end, even though the process was painful for everyone, the South Wellington Special Education Group did make a positive difference. Parent Beth Jackson notes that "We were very fortunate because the school learned from some of its mistakes with other families so our [family's] road was a little easier." Despite what he has been through, superintendent Ammons, is optimistic. He knows there will always be "bumps in the road," but he says, "In any school system I would be in, I would want there to be a special-ed resource group." He pauses, smiles, and then adds, "Although people wouldn't believe that!"

4

CONFLICT ABOUT SCHOOL CHANGE: THE STORY OF SPRINGFIELD REGIONAL HIGH SCHOOL[1]

Where but in small-town America would residents vote in a local election to show their opinions about school practices and their confidence in top school administrators? The names of the superintendent and two administrators at the high school actually appeared on the ballot for individual votes. The June 2000 vote came at the end of the first year of Springfield Regional High School (SRHS) as a school because some parents, unhappy about the education their children were getting in this brand-new school, were able to collect enough signatures on a petition to get the referendum questions on the ballot.

What went wrong? Didn't school officials involve parents and community members in planning the philosophy and curriculum of school before it opened? Were the administrators misguided? incompetent? perhaps too arrogant to listen to parents' concerns? or simply too busy? And what about the school today? How did the vote turn out? What has changed in the school and in the community?

The story of SRHS shows that, despite the best efforts of everyone involved in planning, the change process is complicated—and is likely to be very contentious. It also reveals, though, that there is hope for those who would like to introduce innovative educational practices even in very conservative communities. If their beliefs and commitment remain strong enough and long enough, eventually they can find ways to tame the troubled waters.

PLANNING FOR THE NEW HIGH SCHOOL

For about 30 years there was no high school in Springfield. The town and its two neighboring towns paid tuition for their secondary students to attend a

Sidebar 4.1

Springfield Regional High School Belief Statement

Springfield Regional High School will provide all of its students with equal access to a high quality educational program that will prepare them to meet the demands of a changing world. Our school will be a place of learning for all members of the community. We are all responsible for the quality of that learning. The school's main goals are to meet the varied needs of students, and to foster outstanding achievement and overall excellence. To accomplish this goal, we believe that:

Learning
- Learning should be lifelong.
- The curriculum must be responsive, purposeful and appropriate to the learner.
- The curriculum must be interdisciplinary, integrating relevant information from different subjects.
- Appropriate testing and assessment must measure learning.
- Learning must be active and include solving problems or performing tasks under real life conditions (authentic).

Students
- Students learn through encouragement.
- Students learn in different ways.
- Students must take responsibility for learning.
- Students must achieve learning at high levels.
- Students must have access to a variety of learning opportunities that will serve their needs and interests.
- Students must learn to think critically and achieve their best.

Parents and Community
- Parents are the primary educators of our children. Their involvement is essential to effective education.
- Community involvement enriches learning by increasing resources and services offered to learners.
- The community must give the school its financial, social, and moral support.
- Our community is a part of, and is influenced by, a larger national and global community.

Teachers
- Teachers must believe that all students want to and can learn.
- Teachers must be well qualified in their subject area and effective with various teaching methods.
- Teachers must be lifelong learners.

(continues)

> **Sidebar 4.1** *(continued)*
>
> - Teachers must have personalities that enable them to be effective with students, parents, colleagues, and the community.
> - Teachers must be committed to their students.
> - Teachers must challenge and inspire each student to achieve his or her highest potential.
> - Teachers must provide opportunities for students to be planners, designers, creators, and critical thinkers.
> - Teachers must be viewed as professionals who deserve our support in their growth and development.
> - Teachers must be positive role models for students.
> - Teachers must be able to work in a team.
>
> *The School*
> - The school must be safe.
> - The school must support active learning.
> - The school must recognize individual and team accomplishments.
> - The school must enforce clear rules and expectations for its students and staff.
> - The school building must be attractive and include all indoor and outdoor facilities that we require to meet our goals for education.
> - The school building and grounds must be flexible, open, light, and in harmony with the surroundings.
> - The school must foster good citizenship.
> - The school must use up-to-date technology as a tool to broaden and enrich learning.

high school in a nearby city. When school officials were notified that this school was too crowded to continue accepting tuition students, Springfield school officials had to consider other options. After much discussion and debate about its location, the three towns finally agreed to build their own school. Between the design and construction of the new building and its opening, however, school officials and parents had to scramble to find schools that could accept tuition students for a few years. No one school could accept all of the students. So, when SRHS opened in the fall of 1999, its student body was comprised of students who had previously attended about 12 different schools. And SRHS's first senior class was smaller than it might otherwise have been because some of the potential twelfth graders were "grandfathered," or allowed to remain where they were. This seemingly unimportant fact turned out to be part of the problem that occurred during the school's first year of operation.

Rarely would one find a school that was planned more carefully—or with more community involvement—than was SRHS. Even before the ground was broken for the building, many educators, parents, students, and community members spent a great deal of time working on committees, sponsoring public meetings, visiting other nontraditional public high schools, and listening to experts they brought in as consultants. The superintendent also invited education students from a liberal arts college nearby to assist in doing research projects for the school district for each of the two spring terms before the school opened. The goal of all these efforts was to create a high school whose philosophy and practices would reflect the most recent and best research on student learning. Because its mission statement and core beliefs were similar to the principles of the Coalition of Essential Schools, a national organization committed to helping schools introduce changes. SRHS became a member of this organization.

In 1998–1999, the year before the new school opened, the advance team, a principal and two deans (assistant principals), began working. They were charged with overseeing the process to create everything a totally new school would need—from choosing school colors and a mascot to developing a curriculum framework and student handbook. In the early spring, they also began hiring teachers who would support the innovative plans for the school. But the advance team did not work in a vacuum. Parents, students, and community members were involved in a variety of ways from serving on subcommittees to participating in open discussions at public meetings.

The Curriculum Focus Group representing all three towns, for instance, completed the mission statement and posted it on the Web, noting that "[a]ll of the school's significant actions should support the mission statement." The site included an e-mail link to an advance team administrator so anyone who wished to could comment on the draft. The mission statement read:

The Springfield Regional High School Mission:
A unified community pledge:

- To teach all students to use their minds well and to cultivate their particular talents
- To establish a culture of respect, responsibility, service, and courage
- To demand excellence and to foster lifelong learning in a safe and welcoming environment

The advance team did all it could to make sure people knew what was happening with the new school. There were newspaper articles and open meetings. The mission statement provided a clear framework for the philosophy that would be the foundation of the culture and structure of the school. SRHS would be a very different school from the idea of high school most people have—and quite unlike the mostly traditional high schools future SRHS students were then attending as tuition students. Everyone hoped—and school officials believed—that the new school would provide an engaging academic curriculum and a caring environment that would promote greater student achievement and success.

WHAT SRHS DID DIFFERENTLY

SRHS differs from most high schools in many ways, but several practices were central to its mission and core beliefs—and some of these created a more negative response from parents than others. If some of these practices sound confusing in the following explanations, readers may begin to see why parents later became very upset.

Unlike most high schools, SRHS classes are not tracked; students are grouped heterogeneously. Ninth and tenth graders form teams of about 80 students and four teachers with whom they stay for two years. Although there are no honors classes, students may choose to participate in Honors Challenge, assignments within a class that deal with the content in more depth and breadth.

Because teachers and students work in teams, SRHS can focus on more interdisciplinary learning. In fact, students in grades 9 and 10 do not even take English and Social Studies. A double-period course called Humanities is taught by two teachers. Even when the 40 students assigned to each block meet in separate classes (20 with the "English" teacher, 20 with the "Social Studies" teacher) rather than as a class of 40 altogether, their work has been planned as a unit by both teachers to address specific essential questions. For example, one tenth- grade class focused on the questions "What is a revolution? What is progress?" Part of the time they studied the American Revolution with the social studies teacher; the rest of the time they were exploring these questions with the English teacher in a different way as they read Alice Walker's novel *The Color Purple.* The school schedule includes time for team meetings and planning.

The curriculum at SRHS is entirely standards based. "Seat time" and apparent effort will not substitute for performance. Instead, students need to demonstrate that they have met the learning standards set for a particular

Sidebar 4.2

SRHS General Assessment Rubric

	Competent	Advanced	Distinguished
Level of understanding	Student work demonstrates a command of the identified knowledge and skills of the course.	*In addition:* Student work demonstrates a greater breadth of essential knowledge and skills and a deeper understanding of identified concepts and skills.	*In addition:* Student work demonstrates a clear and integrated understanding of identified concepts and skills, and/or student demonstrates development of new understanding.
Application	Student work demonstrates concrete connection among ideas. Student work reflects the recognition of potential problems and questions and a basic understanding of how to deal with these issues. Communication is direct and concise and approaches the detail necessary to convey an in-depth understanding of concepts.	*In addition:* Student work demonstrates complex and diverse connection among ideas. Student work reflects careful consideration of problems and questions. Communication is effective, varied, and detailed.	*In addition:* Student work demonstrates an intellectually advanced connection among ideas, including abstract connections. Student work demonstrates a process by which problems and questions are resolved.
Quality of work	Student work meets the identified criteria for task completion.	*In addition:* Student work elaborates and extends on identified requirements in a clear and meaningful way.	*In addition:* Student work elaborates and extends on the identified requirements, synthesizing relevant concepts.
Date of completion	Most of the work must be accepted and completed in a timely manner.	*In addition:* Almost all student work must be attempted and completed in a timely manner.	*In addition:* All student work must be attempted and completed in a timely manner.

INC—Incomplete
For a student to receive an "INC," most of the work is attempted but some of the outcomes of the course have not been met at the competent level. Competency will be demonstrated within the specified time.

NC—No Credit
Student did not attempt a significant amount of work or did not achieve a significant percentage of the standards. A zero will be averaged into the student's GPA, and the course will appear on the permanent transcript.

Sidebar 4.3

Grading and Reporting at Springfield Regional High School

Standards-based Grading
Under this system, teacher start by establishing the essential skills and knowledge that students must achieve in order to receive credit. Final recorded grades will demonstrate the degree to which students have mastered these skills and knowledge.

A Basic Tool and How It Works: Clear and Public Rubrics
A rubric is a tool that describes how achievement "looks" at each level in terms of clear, commonly understood criteria. Teachers will use rubrics for each major assignment, so that students and their families understand expectations as well as final grading decisions. The general rubric here defines the criteria that are held in common by all classroom teachers for overall course achievement. Each teacher will "tailor" this general rubric to meet the assessment needs of particular courses.

Performance Levels
Achievement will be recorded on transcripts in terms of the following performance levels:

    Distinguished (DS-, DS, DS+)Incomplete (I)
    Advanced (AD-, AD, AD+)No Credit (NC)
    Competent (CO-, CO, CO+)

The general rubric describes in broad terms what each achievement level will look like.

Honors Challenge
The Honors Challenge is an assignment or a group of assignments in which participating students explore course material in greater depth and breadth. Students who successfully complete each part of the challenge at the Advanced level or higher earn an Honor designation on their transcript.

Reporting Schedule
Narrative Progress Reports in Each Course: midquarter
    Student-Parent-Advisor Conference: the end of the 1st and 3rd quarters
    Report Cards: the end of each semester
    In the spring, seniors and sophomores will replace the conferences with exhibition of work: the Senior Celebration and Sophomore Core Portfolio, respectively.

Honor Roll
Honor Roll: all courses in the Advanced range or higher
    High Honor Roll: all course grades in the Distinguished range

---

(continues)

Sidebar 4.3 *(continued)*

Academic Index
SRHS uses the following Academic Index to report overall academic standing on transcripts to colleges and employers.

DS+ = 9   AD+ = 6   CO+ = 3   NC = 0
DS = 8   AD = 5   CO = 2
DS- = 7   AD- = 4   CO- = 1

Transcripts indicate whether a student's academic index was in the top 10, 20, 30, 40, or 50 percent of his or her class and will include a translation to a 4.0 scale.

---

course in order to earn credit. The school has a general assessment rubric that teachers must use for all major assignments—although they may "tailor" it to suit their needs in a particular course. (See sidebar 4.2.) Grading is also standards based—students' work is assessed as *competent, advanced,* or *distinguished,* according to the criteria spelled out in the school's general rubric. Students not achieving competence in a course do not fail. Instead their work is designated as "incomplete" or "no credit." To earn credit for graduation, they must make up the work (at the competent level), perhaps by attending summer school. SRHS already has in place a curriculum based on the state learning standards, which a recent state law mandates that all school districts have in the near future.

Because the grading system was different, SRHS thought parents and students would understand it better in terms of traditional grades. Thus, *competent* was compared (roughly) to a C; *advanced,* to a B; and *distinguished,* to an A. Rather than increasing understanding, however, this approach turned out to be very problematic. So, in the middle of the first year the school board voted to discontinue the traditional grade equivalents and to revise the process for determining grade point averages (GPAs). GPAs are now figured on a 9-point basis, competent (CO- = 1 point, CO = 2 points, CO+ = 3 points) on up to distinguished (DS- = 7 points, DS = 8 points, DS+ = 9 points). This index is provided on transcripts when students' overall standing is reported to colleges and employers. Grades are reported to parents eight times a year. Four narrative progress reports are sent at midquarter. At the end of the grading periods there are two report cards and two student-led conferences, one in the fall and one in the spring.

Another important part of every SRHS student's experience is the Roundtable. Roundtable teachers function in some ways like the better-known "teacher advisors" many other schools have, but at Springfield Roundtable also includes some academic work. The Roundtable, a small group of 10 to 12 students, meets with an adult every day for a half-hour to discuss students' concerns and take care of some matters in the way the typical homeroom might. But Roundtable teachers are also advisors and advocates for their students. They maintain contact with other teachers about their students' performance and communicate with parents. And they also have the responsibility to supervise students' work, as tenth graders, on the required portfolio and, as twelfth graders, on the required "Senior Celebration," an independent student project every senior must complete successfully in order to graduate. Thus, the Roundtable is not just a way to ensure that every student has a close relationship with at least one adult in the school but is also an essential part of the SRHS curriculum.

When SRHS opened the doors to the first students in September 1999, everything was new! The building is spacious and attractive. The main entrance opens into a well-lit, two-story lobby with windowed walls looking into the media center and the main office. The school, with a separate wing to house seventh and eighth graders, features a welcoming cafeteria with a few coffee shop–type booths; high-tech computer and media facilities and equipment; a large gym; well-equipped classrooms; and brand-new books. The faculty, recruited from many other schools, had just begun to work with each other, and most were also new to the community. The entering students, coming from 12 different schools, brought with them their own memories of those schools and different ideas of what "school" should be. Although the three administrators had had the benefit of a year of advance planning, even they did not know what to expect. Believing in a mission for a start-up school is not the same as putting these ideals into practice. But here they all were—administrators, teachers, students—embarked on a new adventure together. They would create the school culture. They—and parents and community members—would begin making the history of the school.

SRHS BEGINS ITS FIRST YEAR

With so much to do and so many changes for everyone to get used to all at once, the first year was somewhat chaotic. Administrators admit that they didn't think too much about parents' concerns early on. As dean of faculty Dennis Perkins explains:

> We had an inkling [that parents were unhappy] before we opened. We had a whole year when we were planning before the school opened. And we did several parent forums where we did PowerPoint presentations and ... a couple of Q and As [question-and-answer sessions] about what we were planning to do.... [But we] didn't really develop the grading system until the summer before we opened so there was no clue about [parent reaction to] that because we hadn't dealt with it.... We knew we were going to be untracked. I was hired under that belief. That was something we certainly encountered resistance to from some of the same parents who led the charge against it once we did open.

The parents were invited to participate in student-led conferences in the fall, which, says Perkins, were very well attended. "They were great. They were the first major success." But, he continues, "that was the only parent thing from November until we were told we had to have a meeting [when] there was a demand that we have one at a school board meeting. Carissa Curtis [a parent] stood up and said, 'I represent 25 parents and we demand that we have a forum to address some of our concerns.'" That meeting, however, didn't happen until February, so, for several months, most of what parents knew about the school came from what their children said at home.

Both administrators and teachers were well aware that some students were not happy. They didn't like the standards-based curriculum and grading practices because they didn't understand them. This approach was also new to teachers. And since they were trying to figure out how to make it work, teachers weren't in a good position to clearly explain the new approach to students. As one student remembers, "Last year I had one class where I think even the teacher didn't know what was going on, but this year the teachers have done it for more than one year and they know what they're doing. They're more organized and have more things planned out for us."

The standards-based grading system in particular seemed unfair to students because the original version attempted to show a correlation between A, B, and C and SRHS's competent, advanced, and distinguished. Perkins explains why midyear, the school stopped using the comparison to traditional grades: "We did a disservice to kids in how it looked.... Because it was harder to get distinguished here than an A elsewhere and we were making them equivalent, it looked on the report card like kids were working harder and getting lower grades. Getting rid of that correlation altogether helped us report who was failing more equitably."

Because no students had been in this school before, there were no older students to acclimate younger ones to the culture of the school. Typically only the new ninth graders in a school need to be socialized, not the entire student body. And the senior class, smaller than future senior classes would be because some students were allowed to finish at the schools they had been attending, was not much help. The ones who came had a difficult time. Perkins says:

> They in particular struggled—in part because the whole mystique of senior year nationwide is that it's a blow-off year, a year of coronation, not a year of working hard. We had this senior celebration, the big senior project they had to do in order to graduate, and we stuck to that. So they went kicking and screaming the whole way. So they were probably the most stubborn. It was also a difficult thing because you rely on your senior class to provide leadership—at least some segment of your senior class—and this was a group that had no experience with leadership because, by and large, the kids who were successful stayed where they were. They had to deal with all this change so they were even more frustrated . . . than they were in their old school. That was a real tricky piece for us.

Perkins notes that what happened was similar to what first-year college students experience. "One of our guidance counselors said, 'Any time you have kids go to a new school, their grades typically go down until they adjust to the new standard.' And we had a situation where every kid in the school was in that situation. . . . I think we had higher standards than many of the kids were accustomed to, and they had lower grades . . . than they were accustomed to." He could see where students were coming from:

> If you are working harder and getting lower grades, that's pretty much a recipe for student dissatisfaction. That was at the core of a lot of the kids' resistance early on. If the kids are unhappy, then the parents are unhappy. My kid's coming home mad and frustrated. "I don't understand this grading system. Last year I was getting all A's, and this year I'm getting all *competents*. I don't even know what that means!" That fed a lot of the parent anxiety. The largest was a fear that their kids were going to be ill-served by this system, that they were somehow going to be sacrificed in the name of educational reform. [Parents] didn't understand the merits of what we were doing. All they saw was a possible loss for their kids. . . . We certainly got the most

pressure from the kids' parents who would have been in the highest track class or honors track class.... Sure, ... [these untracked classes] may help some other kids, but how does this help my kid? It doesn't. It's just a loss. And the grading! It is about the most ingrained educational practice there is—A, B, C. That, of all things, people know and understand as universal. When you mess with that, you're bogus. It's like violating some kind of taboo. The parents worried that [our grading system] was going to jeopardize their kids' opportunity to get into good colleges ... because it was different. Again, they saw it all as a potential loss with no potential gain.

As Perkins points out, though, students began to understand and accept what SHRS was doing before parents did:

There was an interesting lag time throughout the year between what students felt and what parents felt. The student anxiety probably peaked in November-December, and the parent anxiety peaked in February-March.... Sometimes parents were dealing with issues as if they were static when they were changing all the time. Kids' feelings were changing all the time, as they got more accustomed to this or that, or they began to see the benefits of a standards-based system, or they began not to resent an untracked system so much. Definitely the main source of information was from the kids, but [parents] were sometimes holding kids' anxiety from the fall into the spring, and kids were feeling very differently.

All of this anxiety was voiced by parents at an emotional meeting in February 2000, five months after the school opened. Hindsight is always 20/20. Perkins now says, "We needed to have more opportunities for parents to get to know more people in the school. Social stuff. We didn't have enough parent forums in the fall because we were so preoccupied with surviving internally that we didn't look outward. And that was a mistake. We needed to add ice cream socials, fun interactions, or more informational stuff.... [T]here was a bubbling cauldron that we did not address until it was a bubbling cauldron." The cauldron bubbled over in a very public way.

THE WINTER AND SPRING OF DISCONTENT

Because of the directive from the school board, SRHS administrators planned a public meeting for parents, but all didn't go as expected. Dennis Perkins learned from what happened: "We planned the forum and had this elaborate plan and put it forth. [But] Carissa Curtis [the parent who demanded the meeting] stood up and said, 'This isn't what we wanted. Why wasn't I called?' From then on, we have involved parents in the planning."

After she made her request for the meeting, Carissa remembers that "Dennis asked how we would like the agenda set up. I asked that it be an open one to allow parents the opportunity to vent and ask their questions." But, she says, the school plan was different:

> On the night of the parent forum, we had a significant showing of parents. [The local newspaper estimated that 150 people attended.] We were given an agenda that was very different from what was discussed at the school board meeting. Parents, as well as myself, were very upset. So I asked that the agenda be disregarded and an open agenda be allowed as [we had] previously discussed. Dennis and Judith [the principal] refused. We followed their agenda and broke up into groups and met with individual teachers. Unfortunately, this went very badly. The teachers felt attacked, and little was accomplished. All this overflowed over to the students who were also very upset. Parents were contacting the board members; teachers felt they were not being validated for all the hard work they were doing (which they were, and most parents and students agreed with this). Later Monthly Parent Forums were set up; several parents met with Dennis before these took place to discuss how and what would be on the agenda. One of the parents was a professional facilitator so she... conduct[ed] the meetings. This proved to be very effective.

A teacher has a slightly different memory of what happened at the first meeting. When principal Judith Sendak stood to present her speech, the audience responded in a very unexpected way. "We were totally blindsided.... People started standing up and shouting at her. It was very uncivilized. I couldn't believe that people were yelling in front of all of these people at our principal." A parent quoted in a newspaper article said, "There was almost a riot at that first parents' forum. There was a sense of being handled, using process as a way to circumvent how people were feeling." This parent was probably not the only one who had been surprised to learn that SRHS was taking such a nontraditional approach to education. Even though there had been a great deal of advance planning, information, publicity, and several public meetings before the school opened, some parents may not have paid much attention until their children actually were students in the new school.

The parent forum served as an opportunity mainly for parents to vent their negative feelings about the school. The result was that teachers became discouraged and began questioning what they were doing: "Last year I wanted to quit just about every day." "It was crazy. It was mayhem. We didn't know how the grading system was going to work, but we were doing it at the same time." The school was new; teachers had to work hard to do everything for the first time: "We had no plan, no curriculum. We had nothing to go by." The parents' criticism worried teachers. One says, "I had days when I thought it [was] all going to end."

The meeting, painful as it was, seemed to wake up the students. According to one teacher:

> There were some kids, some juniors, who were ringleaders, who were like "Let's get this place. We're gonna bring it down." And when they saw [what happened at] that meeting, [some] kids ... came in the next day, appalled.... Some specifically apologized to faculty members for having led and been a part of it. And I think that that was a huge piece. That was a huge learning for some of them. "Oh, god, look at what I'm a part of!" It wasn't what they had wanted. It wasn't what they had intended. And they saw us the next day shell-shocked, with tears, freaked out completely, and walking through this building, thinking, "I have to get my résumé together again. This is all going to fall apart." I think they were blown away. They had no idea. They were teenagers, and they didn't recognize the impact of their actions.... [So] the next morning they were like "I didn't mean for that.... Wow, I really screwed that up!"

Everyone agreed that the public forum was painful. Emotions ran high. People vented. But this meeting turned out to be only the beginning of a very long and tension-filled season of discontent leading up to the June referendum when people formally expressed their opinions on the ballot.

While teachers and students tried to continue with business as usual, some parents began to organize. They successfully circulated a petition to get two items on the June ballot: (1) a referendum question that combined two aspects of the SRHS approach, the standards-based grading system and the practice of heterogeneous grouping students; and (2) support for (or no confidence in) three top administrators, who were named. Even someone with little knowledge about constructing response items can see that the results to a vote on the first item would not be helpful: How would one know whether a person opposed (or favored) both practices or only one? Nonetheless, the group forged ahead.

At first the conflict seemed to be characterized as "angry parents vs. administrators." The nature of the debate, however, changed when another group of parents formed. Unlike the first group, these parents gave themselves a name, "Citizens for Excellence in Education." A parent leader of the new group and an original member of the school's curriculum committee told a newspaper reporter, "We believe that the Springfield Regional High School is doing a great job carrying out the vision of the community that was expressed in the belief statement." She was so sure of what the school was doing that she also announced that in the fall her son would transfer from a private school to SRHS.

The creation of the Citizens for Excellence in Education was an important turning point in the turmoil. As Dennis Perkins explains:

> They added a countermeasure on the petition. They were articulate spokespeople and gathered support at all the parent forums. There were a lot of things that helped change the nature of the debate, but a key one was [this second group's getting involved]. Initially it was angry parents vs. administrators. And that wasn't good. Then, when we had more parents involved, kids, teachers, and we were facilitating, kind of refereeing the discussion, it was fine. . . . [I]t wasn't us-them. It was parents talking to parents. We could say, "Well, we hear what you're saying, and we agree with that." Or "We hear what you're saying, and although we may not agree with everything, we can work with this concern." It was just a much healthier dialogue, much more constructive. It deflated a lot of that anger because it was parent and parent going at each other instead of all coming at us.

The views of some students had begun to change even before that meeting; others had a change of heart after the first public meeting. Whatever the reason for their conversion, the students helped build support. Perkins says: "Fortunately, by the time the referendum rolled around . . . , the students were feeling better about the school. And we literally had a meeting one time where this parent got up and carried on about how something was horrible about the grading system, and the person's kid turned around and said, 'Mom, I don't feel that way anymore!'"

When teachers and students got involved in the discussions, the issues became more complicated and defused the initial thrust of the parent group with no name. Dennis Perkins explains:

> They [members of the first parent group] were looking for scapegoats, which is why our names were on the petition. They kind of assumed this evil because their assumption of school leadership was that administrators had imposed something on the school and the teachers were just following. The

parents didn't know that the teachers, whom they were increasingly [learning] to like, supported what we did and were very committed to what we did. If they throw out the vision, they are going to throw out all these teachers, too. That was a real big recognition in learning, but that didn't become clear until teachers became more vocal and more involved in the debate. And kids became more involved and said, "I like my teachers. I like what's going on, and don't tick them off." . . . They [the parents] didn't see that it was kind of a package deal. The teachers came for the vision.

Even though the nature of the debate shifted from an attack to a dialogue, the questions on the June ballot were still there. Thus, the school did what it could to educate and build support for what SRHS was doing. Perkins recalls spending an enormous amount of time on public relations. "We sent out a ton of stuff last year. I made up packets—samples of Honors Challenges, samples of rubrics. I spent a ton of my time last year on PR, thinking about ways to get information out, to educate people about what we were doing."

In March the school board voted to discontinue the use of letter grades, thus discouraging people from trying to correlate the SRHS standards-based grading with the traditional A, B, and C approach. The board also endorsed the recommendations made in a report by the Grading Task Force, an ad hoc group of parents, teachers, students, and the school's governance committee. The recommendations of the task force went beyond the consideration of grades and grade-point averages, though, to include suggestions about the Honors Challenge and Roundtable curriculum. All of the suggested changes were ways of clarifying and improving what SRHS was doing to implement the vision. That remained unchanged.

Administrators and teachers gathered information. Teachers wrote letters to the admissions offices of their alma maters about the likely effect of the standards-based grading approach on college admissions. The responses were shared with parents. Admissions officials from area colleges also came to Springfield to reassure parents. They told them that, because college officials were used to dealing with a variety of grading approaches and that grades and class standing were not the only factors used in making decisions about applicants, SRHS's approach would not prevent students from being admitted.

Another resource for SRHS was the principal of a high school in a neighboring state. This school was a member of the Coalition of Essential Schools and had also begun as a start-up school. Because it had served as a model when SRHS was only in the planning stages, some Springfield people had visited the school then. Perkins says, "We talked with [him]. . . . They also had a petition to try to bring them down. They had similar issues, and they survived. That was comforting [to us]. When he came here and spoke, parents and the kids could see that we were not the only place on the planet that was doing this kind of stuff. That helped."

The season of discontent climaxed with the June vote by Springfield citizens, which occurred about the time SRHS was ending its very first school year.

THE RESULTS OF THE JUNE VOTE

According to the newspaper report of the results of the June referendum, both sides claimed victory because people voted both to change things *and* to leave them alone. This confusing result was possible because of the way the two of the questions were worded.

On one question, put on the ballot by the school's critics, 389 voted in favor of having "a traditional, tracked-by-ability program" and a grading system using letters (A, B, C) while only 304 voters supported the school's current practices of untracked classes and a grading system using competent, advanced, and distinguished. Since two different practices had been lumped together, there was no way to tell how people might have voted if these had been separate questions.

In contrast, the vote was 421 in favor and only 255 opposed on another question, asking if voters supported the effort by faculty and administrators to carry out the school's belief statement. This question was put on the ballot by the Citizens for Excellence in Education, parents who supported the school, because they had thought the first question was "misleading and confusing."

The administrators all received votes of confidence in their abilities to implement change at the school. The total vote for the superintendent was 375 to 304; for the principal, Judith Sendak, 391 to 284; and for Dennis Perkins, dean of faculty, 417 to 253.

Just before the referendum, SRHS graduated its first senior class. Speaking to the 39 graduates, parents, teachers, and friends, the valedictorian said, "As the first senior class, we will pave the way for senior classes in years to come." He noted the challenges his class had overcome—and who could forget how tough these were? "We are the better for it. We are better prepared for the future." Calling the students a "courageous and active class," principal

Judith Sendak highlighted the graduation as a "historic occasion" and said, "We have had quite an adventure together."

The first year ended, and summer vacation began.

SRHS: AT THE END OF YEAR TWO

After the tumultuous first year, SRHS opened in the fall of 2000 with a sense of calm. The school with no history the previous September now had one, however brief. The practices that generated conflict in the first year were now institutionalized, understood, and accepted, if not by everyone, then by a majority of students. The school had begun to build trust with parents and community members.

When the new class of ninth graders entered, three classes of older students were there to help teachers and administrators welcome the new students to SRHS. Dennis Perkins notes that the teachers were in a better position, as well:

> Because our teachers had been through last year, they knew better. Just like any teaching—your lesson is going to be better when you can anticipate kids' misconceptions. So having gone through [one year], they could know that [helping students] understand standards-based grading [would be] hard. "It's going to be hard for you to get the fact that, if you fail something, you can't just walk away from it. You're going to have to do it until you get it right." And changing students' work habits so they don't just blow something off but realize that they're going to have to work until they meet the standards, that's a major challenge. But, as a teacher, if I know that's the case, then I can prepare better for it.

Perkins points out that there was change, but the changes weren't in the practices that parents criticized the previous year. Instead, the administrators and faculty worked hard to bring more clarity to the process and understanding to students and parents. And he says, "The other huge thing—that happens only through time—is that we just get better at what we do. I do think we made substantial changes, but they were in the name of helping us do better what we said we were going to do."

SRHS added an interdisciplinary honors seminar at the ninth-grade level, which meant that all ninth-grade honors students could be together at the same time. Tenth graders could choose to participate in an Honors Challenge unit in which they extended and enriched their knowledge of the unit on progress and revolution that all tenth-graders studied. The Honors Challenge meant reading and discussing another book in small groups with other Honors-Challenge students and developing an end-of-the-unit independent project, which some students did collaboratively. The students formally presented these projects in front of panels of adults, some of whom they had not met before. For instance, for several of these presentations, three faculty members from a liberal arts college in the area served as the panel along with the older sister of a student teacher, just back from working in the United Kingdom.

Parent involvement is central to everything SRHS does now. Perkins explains that parents participate in making decisions at every level:

> We have parents at every role in our new governance structure that we added this year. That's a big change. We have the vision keepers, which is the overarching body. That has two parents on it, but only one comes, who's the school board chair. And we have other subcommittees that have parents on them. Carissa [the parent who demanded the open forum the previous year] is on one.... Every proposal for change goes through this process. It starts with the filter group which decides where it goes. Then it goes to two subcommittees that flesh out the proposal and draft and redraft and then it goes to the vision keepers, which approve all proposals. That [vision keepers' committee] includes four kids, six teachers, a couple of parents, and the principal.

SRHS has also scheduled regular meetings for parents this year, both open ones and others with topics, but the turnout has been very low, only "five or six people," Perkins says. "I don't really understand why. And it's certainly something we need to work on next year." Getting more parents involved is important.

> Our feeling is that the more people know about what we do, the more successful we will be and the more support we will have. We want them in [the school] as much as possible. We want them to see student work. We had very well attended student-led conferences in the fall and in the spring. About 90 percent of parents [came] in the fall and the same in the spring. So they are in the building. We have them involved in the senior celebrations and the sophomore portfolio presentations. We want them in seeing their kids' work, meeting teachers, because we know that if their kid's happy, they'll be happy. And the more they interact around their kids' learning and get their concerns addressed right there, the less they will be looking for things structurally that must be at the core of their kids' struggles.

The school got some good press this year. Positive stories about the school and its new approach to education appeared on television and in local papers. One student told a reporter, "[In other schools] it's more of a dictatorship. You did what they told you to do. Here they actually want to hear what you want to say." Another said, "They try to stay away from the textbook way of teaching. It's more hands-on learning." The principal said, "I think it's going better this year. I'm excited by how many kids have become real learners—not just for grades but because they've caught the fire."

Everyone agrees that the culture at SRHS is different from most traditional high schools. "What is working [here] is that kids are validated as whole human beings. They are not verbal scores. They are not mathematical logical scores. They are real kids and they are beginning to buy into that." One student responds to the question "What's working?" by saying "We could be here all day. You know, there's so much stuff. I think the curriculum is great. I think that the grading system is great, and I think that the faculty is great." Another student remarks, "It's a safe environment. People have different opinions but everyone respects everyone else." Students feel very comfortable voicing different opinions, and they agree that they can express themselves freely.

Teachers also like the school environment: "When I walk through this school, I feel very comfortable. I feel welcome by the students and the staff." One teacher shares a story about the "senior prank" in the second year to show how much the kids have learned. Students filled the main office with balloons and wrote on the door, "It looks like you are going to have to be creative and practical problem solvers." (The phrase "creative and practical problem solvers" echoes the school's mission statement and the state's learning standards.) The teacher says: "The community of the school itself has grown and developed a lot in the last couple years among the teachers and the students. The fact that our senior prank was a demonstration of students as creative and practical problem solvers says that we have kids that at least have heard and learned some of what we're about."

If parents are still unhappy, Dennis Perkins hasn't heard from them, even though he keeps asking "Are there burning issues? Let me know. Call me." He adds, "I haven't got any response. It's weird." He thinks the school board members agree that, since board members are "tougher" on the school, parents don't need to be. "The parents filter their stuff through the school board, and the school board raises it in a public forum with [cable]

TV cameras on them. Parent gripes get dealt with more efficiently. I think that's their perception."

Some parents may be taking a wait-and-see approach, as Carissa Curtis is:

> When I first approached the school board, I was for returning to the traditional methods of teaching. As I became more and more involved in the parent forums, and with the committees that were set up to address the topics identified by the parents, I began to have a better understanding of the present system and how it would benefit all students. It was then I decided to support the existing system. Had communication with administration and teachers not broken down early in the process, I believe that much of this could have been avoided. . . . The staff was also brand new and most never had utilized these methods; and the administration not only had the high school to contend with, but also had a brand-new middle school [located in a separate wing of the new building] to deal with, as well. Their dedication to this school is a credit to all three and the faculty, as well.
>
> This year I have remained involved only minimally as the . . . issues are no longer a problem with my son, who was a junior this year. He is the one who has complained that he is not challenged in [some of] the classes that he takes that are heterogeneous. He also verbalizes that the grading is more rigid. His frustration comes from the slower students, and even more so from those that are regularly distracting. . . .
>
> My impression with this system is that it is basically a good one, but [it] still does not meet the needs of those students who not only are advanced but want to be further challenged. My son is one of these students. The school is in the process of creating a junior/senior curriculum to meet this need. . . . I have mixed emotions. . . . I believe that the present system is a good one, and that, given time, experience, and training on the part of students and teachers, it will be here to stay. Had my son not had the exposure to another high school and had he had this experience earlier in middle school, I don't believe there would be any major concerns. The system is not a "perfect" one, but neither is the traditional system. I am reassured by a faculty and administration that continue to address the issues, consult parents, students, and community, and make the necessary adjustments in an effort to accommodate all students at all levels in spite of budget limitations.

Another parent, John Blackburn, whose son is "average," likes the SRHS approach and cites only one minor concern now.

> I was always very open to the new ideas of SRHS and hoped that it would succeed in its stated mission. The pressure was tremendous to return to the traditional, "that's how I learned" system. I really felt that this system would

be the best for my child. He was average. The average student never gets much service in the "traditional" system because the misbehaving and talented kids got most of the attention. This system was going to treat all equally. I was for it. I was really disappointed when the administration felt compelled to try to take a half-step back and bring a form of traditional grading to this environment. I think it actually set them back a year or more. The only other issue that really bothers me is the use of the word competent. I wish there was another word or phrase because it is disappointing to have your son come home and say, "I am incompetent," when a task is not completed. That reverse meaning [of competent] is way too strong. There are other words that could fit the level but not create that feeling.

The second year has gone reasonably smooth for our son. Gone are the references to [his previous school].... There are still some real issues of consistency among the teachers..., [but] my feeling is that it will only get better.

One clear sign that SRHS has turned the corner shows in its increasing enrollment. In addition to the students from the three towns that had built the school, SRHS has been accepting tuition students from Rawson, another town without a high school. This year, however, the enrollment of 804 projected for next fall exceeds the 775 limit for which the school was built. Thus, the school board voted to notify Rawson that no more tuition students would be accepted. Rawson students already attending SHRS, however, will be allowed to stay and graduate.

As SRHS ends its second year, people seem to be thinking of it less as a "new school" and more as "their" school or simply "the" school. Everyone in the school has remained committed to its vision and mission. Even parents who may still have some reservations praise the efforts of administrators and teachers and want to give them a chance. Other educators from other school districts look to SRHS as a model because it has actually implemented all of the changes a state blue-ribbon commission on secondary school reform recommended for all high schools in the state. Because SRHS now has a track record and a presence as the community's school, its future looks much brighter than it did during the dark days of that first year.

THE DEAN OF FACULTY REFLECTS ON THE CHANGE PROCESS

As Dennis Perkins recalls the struggles of that first year, he paints an interesting picture of the change process and gives some advice to other educators who are contemplating changes in traditional schools.

Sidebar 4.4

Springfield Regional High School Governance Structure

Main Features of Our Governance Structure
- Allows for a wide representation of various constituencies at every level of the process
- Allows for flexibility within a structure
- Reflects the values and culture of this school
- Allows for a broad range of involvement in the decision making process

Process for Issues
Step 1. An issue is brought to the filter group: There will be an administrator present at every meeting to help direct the issue to the correct subgroup or directly to administration.

Step 2. Issues are sent out to the various subcommittees. Each subcommittee is headed by a member of administration. Subcommittees make proposals out of the issues and forward them to the vision keepers.

Step 3. Proposals are sent to the vision keepers who make a decision. Both an administrator and school committee member are present at every meeting. No decision can be made without an administrator present.

The makeup of each of these subgroups is as follows:

Filter:
1 administrator
1 student (a junior or senior)
1 community member/parent
Director of maintenance
1 faculty member

Student Involvement and Action:
1 student per roundtable (approximately. 40 students)
1 roundtable advisor per grade (four faculty)
Dean of Students
1 Member of student services (i.e., guidance, social worker, school resource officer)

Culture:
4 faculty
2 staff members
8 students (two per grade level)
1 parent
1 community member
1 administrator
1 member of student services ie guidance, social worker, school resource officer

(continues)

> **Sidebar 4.4** *(continued)*
>
> Teaching and Learning:
> 1 administrator
> 7 faculty (representing a broad spectrum of the faculty)
> 4 students (one from each grade level)
> 2 community members/parent
>
> Vision Keepers:
> 1 administrator
> 7 faculty (representing a broad spectrum of the faculty)
> 4 students (one from each grade level)
> 2 community members/parent
> 1 member of student services (i.e., guidance, social worker, school resource officer)

> [M]ost of what happened last year was inevitable, and most of the success and change in reaction this year was inevitable.... [T]here was so much change last year that it was literally traumatic for kids and parents and teachers.... We could have done anything, it wouldn't really matter. We could have done all the parent forums in the world, but it was just too much to deal with at once. I still think it was the right thing to do.... [We wouldn't be where we are] ... if we had tried to incrementally phase in something like heterogeneous grouping or ... standards-based grading because there's so much resistance.... [T]his year we are so much farther ahead [even] if we do nothing just because we have existed for a year.... [T]here are all these kids who get it now and they can teach the other kids.... [T]here are all these teachers who get it and have examples of things that have worked and [some things are now] just status quo. Suddenly roundtable, which was crazy last year, has this inertia to it because "Oh, that's what we do. We had it last year." There's just less inherent questioning because it exists. It is the same with tracking.

Looking back, Perkins can even see how having those questions on the ballot was beneficial to the school.

> In some ways it really galvanized faculty around the cause.... Had that not happened, we would have lost more teachers. More teachers would have left because there was lots of hair-pulling and "I don't know if I can do it," "I don't know if I can make it," but, once there was a threat from the outside, it was a Cold War analogy, rally-round-the-flag syndrome. Teachers came together, and we lost very few teachers last year.

The faculty was important in helping the administrators survive the first year. "I relied a lot on faculty support. I believe very strongly in Robert Evans's twist on the golden rule, that the faculty has to be as good to each other as they are to their kids. And I don't think that is often the case. But this faculty was really supportive of each other. That really helped in getting us through." The need for this mutual support is still there because of the conflict:

> There are emotional scars, and a paranoia, and a distrust of community, which runs pretty deep with a lot of faculty. That is the cost. It will take a lot of time. The school board doesn't understand that teachers are really nervous about the community's view of them and why we don't want to just share everything we do with them. They don't understand that teachers don't fundamentally trust the school board and elements of the community because they fear we are going to be attacked. If we show any vulnerability, we will be attacked. [W]e always have to be relentlessly positive in what we share because the perception is so skewed the other way.

Perkins offers some advice to others about making changes in school. First, it is important to involve a broad constituency before a decision is made. "Involve as many stakeholders [in the conversation] as possible, as early as possible, so that you can come out of the gate with any kind of proposal having broad-based support and [everyone] . . . together in presenting this new idea."

Building trust is essential. Perkins says: "We didn't have any trust in the community last year for a very good reason. We had no track record and we were trying all these things with the thing that matters most [to parents], their kids. Understandably there wasn't trust." The task of the school is to build that trust in as many ways as possible. He says: "That largely happens through one-on-one interactions. Returning phone calls. Sending out progress reports. . . . The more they are wondering and worrying and not knowing, that's where the problem starts."

Better communication means that the school can address problems before they happen. At SRHS this communication goes beyond sending information in newsletters—although this is important—to making sure that parents see what students are learning. The student-led conferences are excellent in this regard, but Perkins also says, "The more we can get them in to meet our kids, to see our kids' work, the more they are going to trust us and the less they are going to be worried about the next new idea, whatever it is."

Understanding the change process and the likely patterns it falls into was also useful for SRHS educators and would be for people in other schools as they plan changes. Perkins explains:

> I remember reading in Michael Fullan [author of *The New Meaning of Educational Change*] that in any organization . . . the performance level drops as you first implement a change. That's really helpful. You know we're struggling. Obviously there's crisis of confidence with teachers. Are we doing the right thing? Everyone's attacking us. . . . It feels like the right thing, but then it takes seven months to get any progress. [It helps] to recognize the inevitable pitfalls and the downward trends beforehand. You can't just go from here to here. You have to get worse before you get better. So psychology changes are important.

Thinking about psychology helps, too, when administrators deal with critics of a change. Letting people know that the school recognizes their concerns is as essential as involving them in the change process beforehand. Perkins says:

> One thing we didn't do very well was acknowledging concerns, listening to concerns people had and making sure that they understood that we understood where they were coming from. I think we were too quick to defend and respond as opposed to saying "Yes, I understand why you're upset. Your child has lower grades than last year, and he doesn't understand what's going on." Active listening . . . would really help in acknowledging the frustrations everyone has. . . . That would have helped take some of the sting out of the attacks.

Springfield Regional High School had one major advantage that most other schools will not have when they introduce change. Since it was a startup school, it recruited teachers who were committed to its vision for a new approach to educating high-school students. In fact, the advertising in newspapers and on the Web described SRHS as "the School of Your Dreams." As events unfolded, however, some teachers must have thought it had become the school of their nightmares. Yet the Springfield experience can perhaps offer hope to any school that finds itself embroiled in a school-community conflict. SRHS educators needed—and they demonstrated—high levels of persistence and patience. In the end, though, what probably made the most difference in Springfield and would do the same in other communities is that they stayed on course.

But, equally important, the educators learned how to listen to parents' concerns and tried to anticipate and address them. They knew where they needed to go to do what was in the best interests of students, and they headed in that direction. When their faith was tested, however, they knew that parents and community members must join them. As they held on to their vision, they were tossed about in turbulent waters. Gradually, however, others began to understand and share the vision.

SRHS is not there yet, but this squall is over, and the sea is calm. Yet because educators, parents, and students did survive and developed better skills during the storm, they are better prepared for the next time they encounter troubled waters.

Part II

BEYOND PARENT INVOLVEMENT:
CONNECTING HOME,
SCHOOL, AND COMMUNITY

5

CHANGING THE PARADIGM: THE HOME–SCHOOL–COMMUNITY RELATIONSHIP AS SYNERGISTIC

Tradition is what you resort to when you don't have the time or money to do it right.

—Kurt Adler

Old ways of thinking, old formulas, dogmas, ideologies, no matter how cherished or useful in the past, no longer fit the facts. The world that is fast emerging from the clash of new values and new technologies, new geopolitical relationships, new lifestyles, new modes of communication, demands wholly new ideas and analogies, classifications and concepts. We cannot cram the embryonic world of tomorrow into yesterday's conventional cubbyholes.

—Alvin Toffler, *The Third Wave*

I believe that we have only just begun the process of discovering and inventing the new organizational forms that will inhabit the twenty-first century. To be responsible inventors and discoverers, though, we need the courage to let go of the old world, to relinquish most of what we cherished, to abandon our interpretations about what does and doesn't work. As Einstein is often quoted as saying: "No problem can be solved from the same consciousness that created it. We must learn to see the world anew."

—Margaret J. Wheatley, *Leadership and the New Science: Discovering Order in a Chaotic World*

The tradition of seeing home, school, and community as separate—although related—remains strong but, thanks to the work of sociologists, school reformers, educational researchers, and others, that view has begun to change. For many reasons people have begun to realize that parents, educators, and community members all have important roles to play and must be more connected in their efforts to raise educated and caring children.

The shift from conceptualizing the home–school–community relationships as satellitic to a new paradigm in which the connections among the three are seamless and synergistic will take time, effort, and imagination. In addition to changing the old mental model we carry in our heads, we also have to untangle the complicated webs of one-way streets and roads to nowhere that characterize the individual efforts of many different parents, educators, other professionals, school and community volunteers, governmental agencies, community organizations, and the bureaucratic processes developed over many years. We have to find ways to communicate with each other—a huge challenge in the fast-paced, overfilled lives we all lead today. Yet what could be more important than finding ways to marshal all of the resources available and to coordinate efforts to serve all children well?

This chapter first suggests that there are good reasons for rethinking the meaning of *education* for the twenty-first century and for questioning the continued separation of home, school, and community. Then, because changing the paradigm is a developmental process, it describes the transitional phase with examples of what is already being done in some schools and communities. Finally, because no recipe for creating the seamless and synergistic relationships presented as the ideal for the new model exists, the chapter ends with a description of the characteristics of the new paradigm and poses questions as catalysts for thinking differently about the many ways parents, educators, and community members might collaborate to educate all children well. Even if the ideal is never fully realized, it can serve as a guide for future action. The closer we come to it, the more children we will help.

RETHINKING EDUCATION FOR THE TWENTY-FIRST CENTURY

Although people tend to use the words *schooling* and *education* interchangeably, they have very different meanings: Schooling takes place in a building; education happens everywhere. Schooling includes only the responsibilities given to teachers and administrators, but education includes the responsibili-

> **Sidebar 5.1**
>
> ## Ten Trends: Educating Children for a Profoundly Different Future
>
> - For the first time in history, the old will outnumber the young.
> - The country will become a nation of minorities.
> - Social and intellectual capital will become the primary economic values in society.
> - Education will shift from averages to individuals.
> - The Millennial Generation will insist on solutions to accumulated problems and injustices.
> - Continuous improvement and collaboration will replace quick fixes and defense of the status quo.
> - Technology will increase the speed of communication and the pace of advancement of decline.
> - Knowledge creation and breakthrough thinking will stir a new era of enlightenment.
> - Scientific discoveries and societal realities will force difficult ethical choices.
> - Competition will increase as industries and professions intensify their efforts to attract and keep talented people.
>
> Source: Education Research Service report online at www.ers.org/report.htm.

ties of everyone and, by extension, everything that influences what children learn. Excluding the effects of the media from considerations about education, for example, is a mistake: Their influence on children's learning is powerful. Even though individuals and some groups have protested the negative effects of TV or video games, the community as a whole has ignored this important part of children's education. Should it continue to do so?

There is another reason for thinking more about education in contrast to the more limited concept of schooling: Children are not the only ones in the community who need to be educated. Rapid changes in society have made lifelong learning necessary for adults. When factories close, people need to learn new skills. Even if people keep the same jobs, they often need to update their knowledge and skills just to keep up. Some adults still need to learn to read and write. Immigrants from other countries need to learn English. Research shows that older adults are likely to be healthier and happier by continuing to learn; doing so may even prevent the tragedy of being afflicted with Alzheimer's disease. If we thought about connecting all these learners with school-age students, we might improve their education at the same time. Why, for instance, do we expect teenagers to connect with adults in positive ways when we lock them up for six hours a day in one building? Would rowdy, unmotivated students in a high-school class behave the same way if senior citizens and other adults were also students in their classrooms?

Perhaps the most important reason for rethinking what education should mean, however, is apparent in events, such as the Columbine tragedy, and the many stories the media routinely report about young children killing other children or their parents, gang violence, domestic abuse, and other situations that shock us or make us fearful. Yet most policymakers and others intent on improving schools push for learning standards and higher test scores—changes that will do nothing to solve the very real problems of school shootings or hate crimes. These reformers talk about "education," but the policy changes they propose relate only to "schooling." Some leaders, however, do take a different approach. For instance, Thomas P. Carey, Deputy Secretary for Elementary and Secondary Education in the state of Pennsylvania, in a letter to all superintendents in the state, advises:

> An effective school counters violence and its influences by placing school safety as one of its top priorities. A comprehensive school safety approach includes utilizing parents, families and communities as a resource in achieving safe and nurturing environments where students can learn. Studies have shown a direct correlation between student achievement and parent involvement. Students who are successful in school and are attached to their schools and communities are less likely to commit acts of violence.[1]

The debate about the degree to which the school's mission should focus on intellectual development vs. personal and emotional development is not new; nor is it likely to ever be resolved. Current events, however, suggest that perhaps schools need to do both.

Although her work is not well known by the general public, educational philosopher Jane Roland Martin presents a different way of thinking about combining these two missions: She argues that schools should teach both the 3 Rs (intellectual development) and the 3 Cs (care, concern, and connection, or personal/social development). Martin notes that the "most cursory reading of our daily newspapers and weekly magazines reveals that as a nation we still have not learned to love our children as ourselves or to do for them what the times demand."[2] She questions the wisdom of focusing only on teaching the 3 Rs and expands the traditional idea of what should be considered basic knowledge for all students. Citing some of the many ways society is different today from what it used to be—for example, men and women both in the workplace, more single-parent families—she argues for changing our conception of school from that of a "schoolhouse" to one of a "schoolhome."

The notion of a "schoolhouse" suggests that the major purpose of school is preparing students to be future workers and public citizens, what Martin refers to as the "productive" aspects of society. Good teachers have never overlooked personal development, but the public today demands higher standards so that the United States will be able to compete in the global economy. Few argue for finding out how well schools address students' personal needs, and many critics perceive giving students a voice or adopting a multicultural curriculum as evidence that schools are pandering to students by lowering standards and sacrificing rigor for relevance.

Martin, however, believes that schools can and should do both. The "schoolhome" would teach the 3 Rs along with the 3 Cs, which would prepare students for what she calls the "reproductive" aspects of society, such as caring for others, raising children, and maintaining a home. Martin is not suggesting that schools take on the responsibilities of parents and families. Instead, she would like to see school curricula that would educate students for all aspects of life because the future of this society depends on their future successes not only as workers and citizens in the public arena but also as caring, competent parents and good neighbors in private homes and personal relationships.

Children once learned these important skills and values from women in the home. But now, with so many children coming from homes where no one has enough time to teach what Martin calls "the domestic curriculum," schools must help fill this void and work closely with parents in doing so. The domestic curriculum, Martin argues, is essential for both males and females, but a required course in home economics will not suffice—just as a course is not likely to do much to raise students' self-esteem. Instead, this curriculum must be embedded in the daily life of the school. Students learn a great deal from the school environment, practices, and relationships. By emphasizing the 3 Cs, teachers can help students become more sensitive to the perspectives of others and learn important interpersonal skills as they work with each other on activities and projects that actively engage them in learning the 3 Rs.[3] As noted educators Theodore and Nancy Foust Sizer remind us, "Moral education for youth starts with us adults; the lives we lead and thus project; the routines by which we keep our classrooms, schools, and school systems."[4] In a schoolhome, all students would feel valued and respected by adults and their peers, and they, in turn, would value and respect others with whom and from whom they learn. In addition to developing their abilities to reason, they also would learn to deal with their emotions in a healthier way.

When we think about what all children need to be educated well, not merely schooled in academic subjects, the challenge seems overwhelming. Certainly teachers cannot do this for all children, and, for a variety of reasons, such as drug abuse, lack of time, or lack of knowledge, some parents cannot even begin to meet the needs of their own children. Maybe, then, what we need is a conceptualization that goes beyond Martin's idea of a schoolhome to create a "community home" for all of our children. Parents, educators, and others in the community may be able to do for children what none can accomplish alone.

THE IMPORTANCE OF CONNECTEDNESS AND INTERCONNECTNESS

> Nobody asks the [school] custodian or secretary (both of whom probably know more about what the troubled kids are up to than anyone else in the building) to share their observations about kids.
>
> —*Harriet Tyson. "A Load off the Teachers' Backs: Coordinated School Health Programs"*

We know that children do better when they feel connected to the adults in their lives. For example, research on resilient children confirms "the power of connectedness . . . [to] protect against all adolescent health compromising behaviors."[5] In the past, people lived in communities—small towns and urban neighborhoods—where people knew each other and watched out for each other's children. These communities, though not perfect, usually were composed of people who shared similar values and cultural backgrounds. Teachers and parents met in grocery stores and exchanged information. Families were less likely to be torn apart by divorce. But now, for a variety of reasons, such as changing demographics, diverse family structures, the increased mobility of Americans, and the fast and often furious pace of life, these informal connections and shared values are harder to come by. Educational leader

Paul Houston, recognizing these changes, advocates a new way to think about the role of the superintendent: "Educators are fond of pointing out that it takes a village to raise a child. But this begs the crucial question—what does it take to raise a child? We are no longer a country of villages, and the web of support that historically supported families and children is tattered. It must be rewoven. . . . Superintendents of the future must see themselves as village builders. . . . But they must do so by reaching out to connect to the resources of the broader community."[6]

Sociologists define social connectedness or social capital as "the network of norms, obligations, expectations, and trust that forms among people who associate with one another and share common values." They note that, because social connectedness is an essential ingredient for creating schools that work, not having it is a major impediment.[7] In addition, noted sociologists Robert Bellah and Amatai Etzioni speak critically of the path society is taking toward unfettered individualism. They call instead for building communities in which a search for common values provides a counterbalance to an individualism run amuck.[8]

The challenge today is finding a way to re-create this social capital in a rapidly changing world. Many people look to the schools to do this, but social connectedness will remain elusive if we rely on the old bureaucratic forms of parent involvement. As educational researcher Mary Henry suggests, "[T]he time is right for a shift to postbureaucratic structures that are organic, interactive, participative."[9]

The work of theorists and researchers in several disciplines also supports a dramatic shift away from the linear thinking that led to the creation and proliferation of bureaucratic organizational structures that no longer serve our needs. Many scholars in the hard sciences and social sciences advocate adopting an ecological or systems approach, both for conducting basic research and for solving social problems. They argue that everything and everyone in the world is connected to everything and everyone else. This shift also can be seen in economic matters. When reformers call for schools to prepare today's students to work in a global economy, they do so because they believe that the U.S. economy does not function independently but is instead part of a larger, interdependent economic relationship with other countries around the world.

Chaos theorists, whose work can be found in many disciplines, ranging from the natural and physical sciences to sociology and philosophy, explain this nonlinear, complex way of thinking in another way. Because an action or event anywhere will have an effect somewhere else, people should view the world holistically rather than as being composed of separate, independent parts. Something small can have large effects, and what appears to be a random series of events can be connected in unusual ways. There are, in fact, patterns to be found in chaos. The classic example given to explain this point in relation to the weather is known as the "butterfly effect": a butterfly flapping its wings in China can change the weather in New York a month later.[10] So, too, seemingly random events in schools and communities need further study because by doing so we would surely find surprising and useful connections we have not considered before.

Failure to consider these interconnections is unwise—perhaps even foolish. As futurist Alvin Toffler explains: "We . . . [must] search out the hidden connections among events that on the surface seem unrelated. It does little good to forecast the future . . . of the family (even one's own family), if the forecast springs from the premise that everything else will remain unchanged. For nothing *will* remain unchanged. The future is fluid, not frozen. It is constructed by our shifting and changing daily decisions, and each event influences all others."[11] Philosopher Ervin Tazlo goes further than Toffler to warn: "[We are] part of an interconnected system of nature, and unless informed 'generalists' make it their business to develop systematic theories of the patterns of interconnection, our own short-range projects and limited controllabilities may lead us to our own destruction."[12]

This paradigm shift already can be seen in the field of education. Some scholars have argued that we should be looking at schools from a broader, more comprehensive perspective than has typically been the case. When those concerned with making schools more effective adopt an ecological perspective or systems approach, they see that families, schools, and community agencies do not operate in isolation. Instead, as the writers of an Iowa Department of Education handbook on parent involvement explain:

> [E]ach system affects and is affected by other systems. [Dr. Ira J.] Gordon [an educational psychologist] sees the family as a micro-system, or the smallest system, surrounded by a meso-system, which includes the neighborhood, local stores, children's schools, and recreation facilities. The meso-system is enclosed within an eco-system which includes mass media, the local work or job climate, local agencies and the school system. Surrounding all other systems is the macro-system which includes national policies in areas such as economics, social problems, and political climate. Gordon suggests that the

various types of parent participation—decision maker, classroom volunteer, adult learner, paraprofessional, adult educator, teacher of own child—are of equivalent importance. Each type is necessary, and at various times family members may participate in one way or another.[13]

Joyce Epstein, who has done extensive research on families and schools, points out the importance of recognizing the overlapping spheres of influence in children's lives, but she notes this view is not yet shared by everyone: "In some schools there are still educators who say, 'If the family would just do its job, we could do our job.' And there are still families who say, 'I raised this child; now it is your job to educate her.' These words embody the theory of separate spheres of influence." On the other hand, Epstein continues: "Other educators say, 'I cannot do my job without the help of my students' families and the support of the community.' And some parents say, 'I really need to know what is happening in school in order to help my child.' These phrases embody the theory of overlapping spheres of influence."[14]

Many schools are working to create partnerships that bring home, school, and community closer together. Epstein's research often has been central to these efforts. For instance, schools across the country and the National Parent Teachers Association (PTA) have adopted her framework listing six types of parent involvement. It also informs the framework of this book.

Increasing the involvement of others makes sense because schools really can't do the complex job of educating children by themselves. Paul Houston, for instance, advises superintendents to forget about the traditional "killer Bs" (buses, budgets, books, and bonds) and focus instead on mastering the "crucial Cs." These Cs include "things like connection, communication, collaboration, community building, [and] child advocacy."[15] The U.S. Department of Education also recognizes that schools need to involve others to help them help children. The department established the Partnership for Family Involvement in Education (PFIE) "to raise student achievement and improve schools by building alliances among businesses, community organizations, families, and schools by promoting family-school relationships."[16] In fact, federal policy, through Title I of the reauthorized Improving America's Schools Act of 1994, mandates that, if schools and districts wish to continue receiving Title I funds, they form compacts for family–school partnerships to improve student achievement.[17] Yet even though school reformers, policymakers, and scholars emphasize the need for schools, families, and communities to work more closely together, the idea is still more rhetoric than reality in most places at this time.

Public schools were originally community centered, but, for various reasons, this closeness no longer exists—to their detriment, many say. What may be perceived as only a school issue is really something more:

> School issues are especially prone to be treated in isolation from other relevant community concerns, remaining narrowly focused on professional considerations. Debates over the curriculum or school discipline can be badly misdirected. A question about the curriculum may be a surrogate for a question about economic strategies, while a question about discipline in the schools may really be part of a larger question about how to maintain order in the community. Issues that are misframed this way can't be resolved because the stakeholders aren't all at the table. Issues that we might be tempted to see simply as problems within schools need to be reframed to embrace the larger context of community concerns.[18]

If we are to adjust to the new realities of our society and educate all of our children well, people can no longer take the narrow or satellitic view of public education described in chapter 2. Times have changed, and schools must change with them. Educational historian William Cutler, whose book traces the complex and shifting nature of home-school relationships in America over time, points out that sometimes "[I]t is no longer clear who is in charge at home, let alone how to reach them." In addition, households of older adults, many who have no children at all, are becoming a much larger segment of the population, and "[t]hey are less inclined to favor increased levels of state and federal aid for education."[19] The reality, then, is that an isolationist way of thinking about public education, which before was just a barrier to creating more effective public schools, may now threaten their continued existence. Because the relationship of home, school, and community can best be described as symbiotic rather than separate, and because resources are limited, the responsibility for educating the nation's children is both a moral and practical obligation that everyone must share.

THE TRANSITIONAL PHASE: HOME, SCHOOL, AND COMMUNITY COOPERATION AND INTERACTION

Educators and parents in some schools and communities have begun to cooperate and interact with each other because they have recognized that they can do more together than they could alone. The question asked by educa-

Figure 5.1

Transitional Phase

Moving toward the Ideal
Cooperative, Interactive

- Focus on schooling and personal development; some curriculum individualized
- Relationship school-directed, some two-way communication
- Less bureaucratic, more personal
- Reaching out but proactive
- More inclusive
- Cultural and social differences recognized
- Limited sharing of power; token representation of parents
- Parents as allies
- Some community connections (when helpful to school)
- Educators begin to research their own practice but do not typically include parents in this process
- Searching to find common ground

Educators (and some parents) ask: How can parents, community members, and organizations help us do our job better?

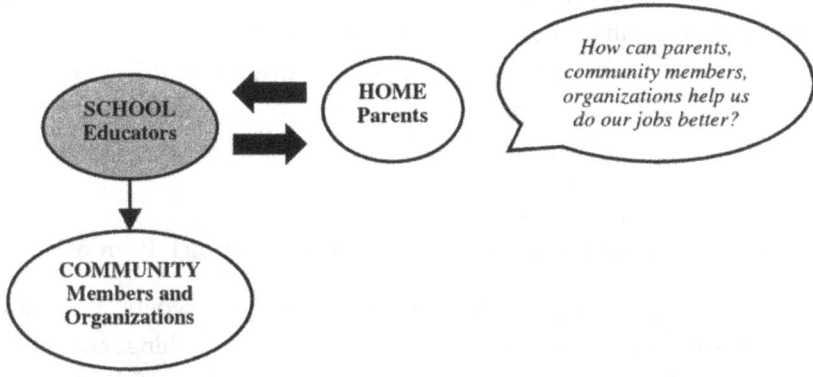

tors (and some parents) has shifted from "What can you do for us?" to "How can parents, community members, and organizations help us do our job better?" A major reason that we characterize these changes as "transitional" is that, while some parents and educators have connected with each other to work for children, other parents may still feel marginalized and the community, although closer in some ways, is still "out there."

The new paradigm requires a radical change in thinking and acting. But changing the old ways, which is difficult for an individual, becomes an even greater challenge for such long-standing institutions as homes and schools. Communities, however, are never static, and a recent event, such as an influx of refugees or a large factory opening or closing, requires community leaders to deal with new circumstances very quickly. Such changes impact schools, forcing educators to adjust current practices in order to meet the needs of children and families unlike those the school had been serving. In such cases, however, while new programs may be added, the institution itself with its deeply imbedded traditions does not change. And even when educators reflect on their beliefs and begin to revise them, changing behavior is difficult. Some view this as being hypocritical—acting in ways that are dissonant with one's stated beliefs. Others, however, describe this stage as being in a state of transition between one belief system and another.

Moving away from the old model of home, school, and community relationships to a new paradigm requires everyone to critically examine the structure and culture of schools and the mission of public education and to commit to developing a new vision. Just as sailors once depended on stars to guide them as they sailed in unfamiliar waters to discover new lands, the model that we present is intended as a guide—not a prescription—for people committed to making schools more effective for all children. The new paradigm is an ideal—not yet a reality that can be mapped with certainty—but, like the North Star once so important to sailors, this ideal points the way for re-visioning home, school, and community relationships.

The Transitional Phase as a Developmental Process

Even though many people inside and outside schools recognize the limitations of the old ways, they have difficulty imagining how things could be different. They should not worry that they have no clear idea of the specifics of the new paradigm. Rather, they should keep its ideals in mind—like their north star—and make changes that they think have the potential of leading them toward the ideal. An old story illustrates the idea of planning in a nonlinear way:

> A stranger comes to town looking for the Brown farm. He stops at a farm stand and asks the farmer, "Where will I find the Brown farm?" The farmer replies, "Just go down the road two Cs and then make a left." Confused, the

stranger asks for more specific directions and gets this response from the farmer: "Go down the road as far as you can see once, as far as you can see twice, and then make a left."

As the story shows, one has to get to a new point before one can see the way to the next one. Because any action has so many unintended results, no one can project where a particular action will lead. Nonlinear planning, which may be uncomfortable for those who like to plan every step in advance, means taking some steps, seeing where they lead, and then planning again.

Changing thinking or changing schools and communities is a developmental process that happens over time. And working with others, especially with people who don't know each other well because their relationships have been distant and formal, takes time, lots of time.

As the chaos theorists suggest, something small can have great effects so the point is to *do something*. We can all learn from experience. A group, for example, can examine any problem, such as reducing the dropout rate in a particular school, brainstorm some possibilities, select the best options, and try them out. Then, after they have had a chance to reflect critically on the results, they can revisit the problem and try again. Through such an ongoing and recursive process, the group will discover what works and what doesn't—and probably will find new strategies to try as they get more information. It's easier to plan the design for a new house by thinking first about what one would change in one's present home and why. But although we could hire an architect with the knowledge and skill we lack to help us complete the design, we can't pay a consultant to tell us how public education should look. No one knows.

Since schools already exist, it is more difficult to think outside the box, to imagine a school radically different from those we are all used to. That's why it is so important to keep focused on the ideals of the new paradigm as we take steps to change the fundamental relationships among parents, teachers, and community members. Because communities themselves must decide on changes for their own schools, these remodeled schools will not look alike. Different communities—rural, urban, suburban—have different needs, different values, different ideas. We don't need "cookie-cutter" schools because a common purpose for public education can be accomplished in diverse ways.

Learning from experience means slowly building on what came before. Small steps are essential in any change process. Consider, for instance, that children do not learn to walk by running in marathons. First, they crawl and later move on to baby steps. Furthermore, children learn to walk without knowing where walking might take them. The climbers who have conquered the summit of Mount Everest, for example, did not even know the mountain was there when they took their very first steps. Baby steps are new beginnings, not endings. Transformative change that lasts evolves over time, step by step.

Many educators, recognizing the problems with the old, satellitic paradigm, have taken baby steps to develop more positive relationships with parents and to establish more productive connections to their communities. At least three factors help explain why they have done so. First, the rebellions of the 1960s led to the erosion of trust in all institutions, including the educational establishment. Second, the changing face of society has resulted in more teachers teaching "other people's children," children from cultures different from their own.[20] For example, even in Maine, the "whitest" state in the nation according to the 2000 U.S. Census, teachers in some communities now have many children from Somalia, Cambodia, and other countries in their classes. Finally, the federal government's focus on compensatory education for children from disadvantaged families and the 1975 legislation requiring the education of children with special needs in the least restrictive environment forced educators to recognize that children are not all the same. The special education legislation, PL 94–142, also mandated that parents participate in making decisions about their children's education.

Educators, however, still believe they have important professional knowledge that others lack. Even as changing circumstances compel them to change, they continue to initiate and direct the home–school–community connections. On the other hand, they have developed a new respect for family differences and community values. The problem is that while they reach out to parents on a more personal level, institutional policies and procedures too often continue to dominate relationships. For instance, when the parents of a child with special needs question educators' decision about a placement, the bureaucratic process they must follow for an appeal puts them in an adversarial relationship with educators, as the story of the South Wellington special-education conflict in chapter 3 shows.

The need to create caring communities for children is great, but, in the transitional phase, doing so means struggling against behemoth bureaucracies. Educator William Ayers describes the dilemma for educators this way:

> Even though we long for community—for places of common vision, shared purpose, cooperative effort, and personal fulfillment within collective com-

mitment—we most often settle for institutions. That is, we generally find ourselves in impersonal places characterized by interchangeable parts, hierarchy, competition, and layers of supervision. Communities have problems and possibilities; schools and universities have departments. We are reduced to bureaucrats, and our sense of purpose and agency is diminished.[21]

The kind of community Ayers believes teachers desire is one everyone would welcome. We cannot "settle." We should not "settle." Every little step educators, parents, and community members take to work together for all children may be a tiny one. But, taken together, these steps can slowly chip away at the bureaucratic stone. Chaos theory again: Small actions can create large effects.

In some schools this process has already begun. The following brief examples show that the transition to a new paradigm is under way especially with regard to parents. The community in the transitional phase, however, is still "out there," although perhaps not as far away as it used to be. (Chapter 9 includes more detailed descriptions of promising practices, programs, and suggestions for moving closer to the ideal.)

Families and Schools

Educators in the transitional phase show that they understand parents to be teachers of a variety of family curricula and values. Individual children come to school with different knowledge and experiences, depending on their families' culture, class, race, ethnicity, sexual orientation, and other factors. The school tries to honor and support parent teachings by diversifying the school curriculum. They take the advice of researcher Emily Style that the curriculum should be both a "mirror" and a "window": Children need to see themselves and their families in the curriculum (mirrors) as well as learn about others who are not like them (windows).[22] For instance, teachers make sure that students learn about the contributions of African Americans and "dead white males" when they study American history. Or they include stories by male and female authors from a variety of backgrounds in the literature class.

Teachers also plan activities to help students learn about each other, often by using their personal experiences as the basis for writing. They create classrooms in which students feel comfortable enough to share their work—their personal stories—with other students. In this way, the children themselves actually "create" curriculum for their classmates to learn.

Teachers call on parents to be teachers by designing homework assignments that build on parents' special knowledge. Students studying history, for example, not only read the textbook, but they also interview parents or grandparents about their lives decades ago. Teachers sometimes invite parents to be teachers in their classrooms, too, by sharing the stories and traditions of their cultures or native countries.

In the transitional phase teachers respect the values of all families and the role of parents as important partners in teaching children.

Home-School Communication

The school no longer depends only on one-way communication. Because what parents know or think is important, educators create opportunities for two-way communication. As researchers Janet Atkin and John Bastiani point out, there are good reasons for listening to parents:

> Listening to parents . . . needs to be seen as a crucial element in any attempt to improve home/school relations. . . . It can make schools aware of the families of their pupils and of the communities in which they are located, more sensitive to the need to consult and share and, possibly, more likely to seek parental cooperation, participation and involvement. Most of all, however, "listening to parents" is a process that can, in itself, identify important educational knowledge and other resources, areas of actual and potential support, and suggest practical developments and new ways of working.[23]

Some one-way communication, of course, is necessary to provide information to parents or get signatures on required forms. But even in these traditional communications there is a change. Because parents from many different backgrounds are the audience for the school newsletter, the field-trip permission form, or the explanation of a curricular change, educators make sure these are "reader-friendly." They avoid educational jargon; they write clearly, simply, briefly; and, if necessary in a particular community, they provide Spanish or other-language versions.

Educators also invite parents to share their perspectives in large-group meetings organized to foster an open exchange of ideas or in focus groups to get their ideas about specific topics. And when parents come to these meetings, teachers show the same respect for them as in the written communications. They avoid "educationalese" and talking *at* parents. They offer food and friendly settings so people feel comfortable speaking their minds. In

Sidebar 5.2

Scott Balicki, first-year teacher
Boston (MA) Latin School

A Letter to Parents

Dear Parent/Guardian,

Welcome to the beginning of the 2001–2002 school year at Boston Latin. As a new teacher at BLS, I would like to describe to you both the Chemistry I curriculum and the manner in which I will be conducting my classroom. I do this because it is my belief that learning is not a process isolated to the classroom, rather one that goes on continuously throughout students' lives. In conjunction with this belief, I do not see parents and teachers as separate or competing entities, but rather as partners mutually contributing toward a common goal: the best possible education for all students. I aim for us to construct and maintain a positive relationship between students, teacher, and parents throughout the year, to build toward this common goal.

In light of this objective, I hold high expectations for all Chemistry I students this upcoming year. It is my intention that all students not only perform well on exams and assignments, but also participate in class discussions, demonstrations, and laboratory exercises. Students can learn in many ways; therefore, I will be using multiple assessment methods over the course of the year to ensure that every student can demonstrate learning and comprehension. Organization will be very important for every Chemistry I student. Maintaining a high level of organization will enable students to function and navigate through a complicated and integrated discipline.

Though challenging, I believe that learning chemistry should be engaging. I construct my lessons to spark student interest in the subject, and put course topics in a modern context. Chemistry itself as a wonderful science; the fundamental study of matter can explain much of the action and structure of our world. Beyond fundamental concepts and calculations, students will explore the history of the major discoveries that shape our understanding of science, as well as examine the impact of modern chemistry on our world today. Students will integrate and apply course work and concepts in the context of modern technology, giving them strong preparation for future academic endeavors in any field.

In order to ensure that you have viewed both this introductory letter and the attached course overview, please sign the attached bottom slip and have your son or daughter return it to me in a timely manner. If you have any questions or concerns, please do not hesitate to contact me at _____ or e-mail me at _____. My office hours will be _____ and _____ and also by appointment. I look forward to meeting all of you throughout the year.

Sincerely,
Scott Balicki

I have reviewed the class guidelines for Sophomore Chemistry I. I understand the expectations for my son/daughter and what I can do to help them be successful. I have been informed of how and when to contact Scott Balicki for consultation.

Name _____ Date _____

some cases, they may hold the meetings somewhere other than in the school building—a public library's conference room or in someone's home—because they know some parents, whose own school experience may be a painful memory, feel ill at ease in school buildings.

Educators know that they can learn from parents so they ask for parent feedback on surveys. They follow up the surveys with focus groups or interviews to get an in-depth perspective on particular issues because some research suggests that responses to surveys may be misleading. For example, one parent who first told the interviewer that she was adamantly opposed to cooperative learning actually believed students sometimes could learn more from other students than from the teacher. Her initial response had been negative because her daughter had recently received a low grade on a group project when the other members did not do their share. Without the follow-up questions possible in a conversation, no one would have known that her first response was inaccurate.[24]

Parents also become important advisors for educators, who actually use their input. In some places, parents share in making decisions. Parents who are members of a school council in Chicago have the power to hire—or fire—the school principal. Unfortunately, such a powerful role for parents is, at this time, still more the exception than the rule. And not everyone agrees that such a role is a good idea. Some argue that the bureaucratic nature of these advisory councils just replaces the current bureaucratic decision-making practices without changing the culture or climate of the school from an adversarial to collaborative one. Yet despite these objections and the fact that some parents do not want to be involved in decision making themselves, most welcome the idea that parents as a group are represented when decisions about school policies and resource allocations are made.

Parents in Schools

Andy Hargreaves, who has written extensively about educational reform, suggests that parent conferences and other communications between the school and the home should be thought of as "teachable moments." He argues that schools and classrooms need to be opened up to parents and community members "allowing learning to run authentically in *both* directions" because it is one of the best ways to build support for public education. One indicator of success in this regard would be seen "when teachers treat every parent/teacher meeting, every child's report card, every piece of homework, and every conversation at the school gates as a 'teachable moment' that they can use to help parents be engaged in and influenced by the work of learning and teaching."[25] Joelle Vanderall, a teacher who never misses an opportunity to connect with parents, says, "If a parent is standing in my doorway, whether they have an appointment or not, I invite them in to sit down and watch what's going on in the classroom, because it is critical to convey the message to the parents that they are a part of the child's success."[26]

Schools in the transitional phase have begun to move in the direction Hargreaves suggests. Although they still have traditional parent conferences and open houses, educators also give parents the opportunity to learn more about their children's progress and participate more directly in their learning in other ways. Many schools, for example, have instituted the student-led portfolio conference: The children show parents samples of their work and explain what they have learned. But even in the parent-teacher conference, teachers make sure parents have time to ask questions and say whatever is on their minds. In the process both the teacher and the parent learn from each other.

Parents, of course, can also attend school events, such as athletic contests and art exhibits, but the school takes extra steps to invite parents to attend and makes them feel welcomed when they come. For instance, in communities in which there are parents whose first language is not English, the announcements that go home and the "welcome" signs at the entrance to the school also appear in other languages. Especially in elementary schools, parent volunteers are important. But in the transitional phase, educators no longer see these parents as merely providing support *for* the school. Instead, they view volunteers as working *with* the school, making it a better place for children to learn. Parent volunteers can be teachers, too, as noted earlier, when they share expertise related to their work, heritage, community resources, community values or needs with students and teachers.

A few schools even have included parents as participants in workshops and inservice programs along with teachers. As chapter 3 shows, conflict between parents of children with special needs and educators in the South Wellington School District led the director of special education to invite parents to attend in-service programs she planned for the special-education teachers. This practice is also a good way for schools to build parents' support for change because parents—not just teachers—need time and opportunities to learn how and why a new practice or program will benefit their children.

Home, School, and Community Connections

Many schools have begun to build better relationships with their communities in several ways that both inform the community and help students learn in more relevant ways. These new connections invariably build support for schools at budget time because more citizens, even those without children in school, cast their votes with direct knowledge of the good work that is being done in local schools. Students' work may be displayed in public buildings, such as banks and community centers, or published in newsletters, which are mailed to parents but are available in libraries for the general public as well. Students also depend on community resources for completing some school assignments, as, for example, when they do short internships as part of a project on career possibilities or when local business and professional people, who do not have children in school, come to the school to do "Career Day" presentations and talk with students.

Schools, especially those in or near urban centers, have the advantage of a multitude of community resources to enrich their curriculum. Some teachers take advantage of the many museums, theaters, symphonies, newspapers, and businesses that have educational programs. Rural communities also have unique ways of inviting schools into the community: Students get valuable firsthand experiences when they visit working farms, cider mills, and other local businesses.

Perhaps the most effective connections between students and the community come as a result of what educators call "service learning." Service learning accomplishes two goals: (1) it makes the curriculum more relevant for students by engaging them in developing knowledge and skills while they work on real projects and (2) the students' work serves a real community need. True service learning goes well beyond the "service requirement" some schools have adopted as a new graduation requirement. While there are certainly benefits for students doing volunteer work in the community, the service requirement is an individual effort not usually connected with the school curriculum. For example, some schools require only that students document a certain number of volunteer hours. True service learning, on the other hand, is always tied directly to what students are learning in their classes. For instance, students in a science class might test water quality in a local water source. Not only do they apply science knowledge and methods in a real-life setting, they provide the results of their work to local officials, who can use the information in solving a community problem.

Another useful connection between schools and the community involves adults: Community members as well as parents serve as volunteers in many schools. Although the number of parents who have the time to volunteer in schools has decreased as more women have joined the workforce, the pool of retired people continues to grow. Some schools have begun to ask senior citizens to help. When older people serve as tutors or mentors for children, they get personal benefits, as well. For instance, one study reported that "volunteering has a protective effect on the mortality" of volunteers; another report indicates that "older volunteers experienced greater increases in life satisfaction over time . . . than did younger adult volunteers." The report noted also that "older adults experienced greater positive changes in their perceived health" than the younger volunteers.[27]

Volunteer hours also can be seen as a way of increasing the school's resources when school funding is tight. According to one report, the value of volunteer time is now $15.39 an hour; viewed another way, in 2000, "[t]he volunteer work force represented the equivalent of more than 9 million full-time employees at a value of $225.9 billion."[28] Even though all of the 109 million people reporting did not volunteer in schools, a good many of them did. And educators probably could get more people to help in schools if they asked them—and made them feel welcome and valued when they came.

Others in the community also can help the schools. Researcher Erwin Flaxman, noting that "[b]y their very nature, schools, homes, and communities are different social organizations . . . [with] complex social arrangements," points out that within these communities are professionals the Annie E. Casey Foundation refers to as "natural helpers." These people, ranging from day-care providers to community health workers, are valuable to the schools precisely because they are part of and have a stake in the community. Moreover, they "understand their neighborhoods and are perceived of as role models in the community." Flaxman, however, advises these helpers to be aware of the language they use with parents and families to make sure both are talking about the same things—a critical element in establishing relationships—because "it would be unfortunate if community members continue to believe that schools and other agencies are just impersonal bureaucracies with useless and arcane rules."[29]

Another way schools have tried to involve parents and community members is to invite them to come together to discuss what students should know and be able to do when they graduate from high school. Using this format, many communities have created what they call "vision statements." The idea is that such a statement, once developed, represents the shared vision of the education people in a particular community have for their public schools and can be used by educators, administrators, and school board members as a guide for action.

Engaging people from the community in a dialogue about education is helpful—and far better than doing nothing. People do learn from sharing their ideas and perspectives; some may even develop ongoing relationships with people they didn't know before. But these "visioning sessions" may be problematic. First, as Peter Senge, an expert on management and the author of *The Fifth Discipline*, whose ideas have been adopted by many educators, warns:

> Actually having shared visions is so profoundly different from writing a vision statement that it's really night and day. It takes a *long* time, and [developing a shared vision] is . . . a process that involves a great deal of listening and mutual understanding. It always involves those two dimensions. . . . The problem is that usually it's *not* a process; it's an event. We all go off and write a vision statement and then go back to work. It's absolutely pointless; it can even be counterproductive because people think, "we've done the vision stuff, and it didn't make any difference." For anybody really serious in this work, you'll spend 20 to 40 percent of your time—forever—continually working on getting people to reflect on and articulate what it is they're really trying to create. It's never ending.[30]

Another problem with visioning sessions is that educators who want to make changes in schools sometimes assume that, because people came to a consensus about the vision statement, they can point to it as the basis for implementing new programs and practices. The problem here is that vision statements are general: "Students will be critical thinkers and problem-solvers." Who would disagree with this outcome? But what the outcome means is not so clear, for different people have very different ideas about what students and teachers would be doing to develop skills in critical thinking and problem-solving. The same words, spoken by different individuals, may reflect very different meanings.

What happens when teachers, trying to address the outcome just given as an example, decide to have students "think critically" about a controversial issue or "solve problems" by working on an ill-defined problem (just like the ones we encounter in the real world) in small groups with classmates of varying backgrounds and abilities? In one school ultra-conservative parents objected to an assignment in a social studies class. Students were asked to choose and present both sides of the referendum questions on the ballot before the November election. These parents did not want their children to hear anything about the positions they personally opposed!

Similar problems are likely in any school for one simple reason: People may agree on broad outcomes, but they usually have very different ideas about how to get to that outcome—and there are multiple ways to work toward any general outcome. The devil is in the details, but *how* educators intend to translate the vision into school programs and practices was never part of the discussion in the visioning session. Thus, educators who engage people in developing vision or mission statements need to see it as only the first step in an ongoing process of translating the vision into practice. Parents and community members need to be included not just in the beginning but all along the way.

No one should worry about lack of agreement, however, because complete agreement may not even be a sensible goal in a society so diverse—and individualistic—as that in the United States. We have powerful and important differences that go beyond philosophical perspectives and differences of opinion. These differences raise questions we rarely discuss—about race, social class, and other issues that are personal and painful to those of us who are not white and middleclass. No form of public education will work for all children until we tackle these issues. The goal should not be to fit everyone into the majority's idea of school but instead to make sure that everyone's voice is heard as together we try to figure out how schools can serve all children well. There can be a *common purpose* for which we join together even when there seems to be no *common ground*.

Parents and Community Members as Activists and Advocates

> *Rich and real parental [and community] involvement requires a three-way commitment—to organizing parents, to restructuring schools and communities toward enriched educational and economic outcomes, and to inventing rich visions of educational democracies of difference.*
>
> —Michelle Fine, "[Ap]parent Involvement: Reflections on Parents, Power, and Urban Public Schools"

Relationships in this transitional phase can be characterized as cooperative and interactive. Educators no longer see parents as problems and critics,

Sidebar 5.3

Margie L. Horton,
Newark (NJ), Schools retired principal

One Principal's Story

In the summer of 1971 my family and I were enjoying our vacation at our summer home in Ocean City, New Jersey. I was really looking forward to this time of rest because the school year had been very stressful. You see, Newark was trying to recuperate from the longest teacher strike in its history, 11 weeks. Though the strike itself was in 1970, it had long-term effects on the staff and strained relationships between staff and administration. I was the vice principal of an elementary school in Newark where very few of the teachers walked out. The union was not pleased with this and made it very well known. Anyway, while on vacation, I received a phone call from the superintendent of schools. He wanted me to come to a meeting being held by parents of another elementary school to see if I could give suggestions to them as to what might be done about problems they were having in their school. I went, not quite knowing why, but when the superintendent calls, you go.

When I arrived at that meeting, I found that the parents were a search committee looking for a new principal for the building. That evening I heard horror stories about books being thrown out of the windows and students actually jumping out of the third-floor windows on to the roof of the portable classrooms outside. I could not believe it. Where were the teachers? Where was the principal? This school housed K, 1, and 4–8 grade students. The second- and third-grade students had been moved to another building, which they referred to as "The Annex" and had its own principal. Even with that, the school was very overcrowded. I listened sympathetically and told them that they really needed to find someone who would be able to turn things around. They needed someone who would be a strong presence in the building, establish order and rules of conduct for the students, and provide leadership for and supervision of staff. I wished them luck in their efforts.

The next morning when I went to report my findings to the superintendent, I ran into some of those parents outside of the board office. They asked me where I would be working in September and I told them that I would be returning to my school. They laughed and said, "I don't think so," and walked away. Though puzzled by their response, I went on into the meeting. The superintendent told me that he wanted me to take over the principalship at this elementary school and that this was a mutual decision made by him and the parents. I really didn't want to leave my building, where I had been a teacher, Title I coordinator, and vice principal for almost thirty years. I asked if I could return to my building if I did not like the position of principal and he said that I could. So I became the acting principal for one year.

What a year that was! When I arrived and settled into the building, I found that indeed there were few books. The stories I had heard were true. I got on the phone and called friends who were principals in other buildings to get enough

(continues)

> **Sidebar 5.3** *(continued)*
>
> books to start off the school year. Then came the students. There were forty or more students in many classrooms with two teachers trying to teach. The overcrowded conditions had to be dealt with or no learning was going to take place. Inasmuch as the parents had already established a relationship with the superintendent, I called on the president of our parent organization to help me find spaces so that we could have more reasonable class sizes and allow the teachers to teach our children. She said that she would get the parents together and speak with the superintendent. I directed her not to come back until we had additional space for the students.
>
> The parents went to the board office and when their request for space was not met they staged a sit-in. They took over the superintendent's office and refused to leave until they were given a space so that the children could learn. They were there the entire day and into the evening. They called other parents to bring food in to them because they were afraid that if any of them left the building, they'd never get back in. By the end of the evening the superintendent had made arrangements for us to move into the Boys Club quite a distance away from the school. That was a Friday night. That Saturday and Sunday, the parents, maintenance workers, my teachers, and I worked to prepare the Boys Club for students on Monday morning. Monday morning there were buses waiting at the school to transport the fourth- and fifth-grade students to their new building. A "teacher-to-assist" was appointed to be the on-site administrator, class sizes went down to about 25, and the teachers were supported in their efforts to bring order and learning back to the children who had suffered through this ordeal.
>
> That year I spent a lot of time driving between buildings, but it was definitely worth it to see my children learning. The parents, the staff, and I worked hard to create an environment where the children felt that they were loved, respected, and expected to do their best. At the end of the school year I accepted the position as principal of the school. The superintendent helped us to acquire a Catholic school building, which had been closed and was within walking distance of the main building. The upper grades moved there and it became "Annex II." The parents maintained a visible presence by becoming paraprofessionals in classrooms, and one parent was even hired to be the secretary of the new building.
>
> I can truly say that I enjoyed my time as principal. I always knew that I could call on the parents, and they had confidence in me that I would always act and follow through to provide the best education possible for the children. This is what can be done when everyone works together, but you've got to put the needs of the children first.

which, of course, they might be occasionally, but as allies even though they and parents still tend to focus on their differing self-interests. The fact is, when both groups work together rather than separately, they are more likely to get the results they individually want.

Most important, all parents, no matter what their race, social class, ethnicity, or sexual orientation are not only accepted but valued because diversity enriches the education of children and energizes the community. But when a community considers diversity in real terms—parents from different social classes or races, for instance—it becomes clear that some parents (or community members) have more power than others to influence the decisions that are made both for the school as an institution and for their own children. Other parents, perhaps less well educated or who speak English as a second language, are excluded or silenced. One reason that honest discussions about differential treatment of students are so rare is that administrators who control what gets talked about in public venues, such as parent meetings, fear the emotions and conflict that will likely result when people express themselves about sensitive issues, such as racial bias or homophobia. Yet Michelle Fine argues: "Without an image of parents and educators working across lines of power, class, race, gender, status, and politics, toward democracies of difference, each group is likely to feel they have gotten no hearing, and will default to their respective corners shrouded in private interests and opposition."[31]

Thus, in the transitional phase, some educators and parents are willing to risk some conflict to begin some tough conversations. These conversations, painful as they may be at first, are important for understanding the many differences there can be in perspectives among people of different backgrounds and different educational levels and with different concerns and priorities. But, as John Dewey notes, "Communication is a process of sharing experience till it becomes a common possession. It modifies the disposition of both parties who partake in it."[32]

Sometimes such conversations come about when educators ignore problems that parents, community members, or organizations outside the school believe are important. The noneducators may become activists to bring about needed change. In Berkeley, California, for example, parents of freshmen at Berkeley High School realized that their children were failing: Near the end of the first semester, half of the school's 300 African American ninth graders were not passing English, math, and/or history. They joined together in a group they called Parents of Children of African Descent (PCAD) and went to the principal to discuss their concerns.

Challenged by the principal to develop a plan, the parent group found financial support and worked with educators to create a program for ninth graders who were failing two or more subjects. Students accepted for the alternative learning community inside the high school were taught in small classes by teachers chosen for their commitment to improving the achievement of at-risk students. Assisted by student mentors, the teachers worked during the year and in the summer to help students catch up with their peers and be ready for tenth grade in the fall. Although the program was successful for many students, lack of funding prevented it from continuing after the first year. Even so, the parent group's efforts made a difference because educators created other less expensive, in-school opportunities for these students.[33]

At this point in the transitional phase, conversations about schools and education may not yet be extended to include community members because, although the community is more highly valued by educators for its resources and still depended on for financial and political support, it has only tentative, tenuous connections to home and school. As schools begin to do more to bring the home and school together and establish some ties to the community, they move further away from the traditional, satellitic model and closer to the ideal, synergistic model.

THE SYNERGISTIC MODEL: HOME, SCHOOL, AND COMMUNITY AS INTERDEPENDENT AND COLLABORATIVE

Everyone asks: What can all of us together do to educate all children?

The new paradigm for the relationships among home, school, and community is an ideal. Even though the synergistic model may never be fully realized, the idea itself can guide school and community change. This paradigm is synergistic because the contributions of educators, parents, and community members acting in concert can produce greater results than those of any one acting separately. And the ultimate objective, that all children become successful, competent, and caring adults—is one that everyone shares. What any society does to help its children learn and grow determines its future.

The synergistic model incorporates many of the practices of the transitional phase, but it goes even further to broaden and deepen the connections between educators and parents and casts the net wider to embrace the community. Relationship-building becomes paramount. Both formal and informal ways of working together are recognized as equally important. Parents, educators, and community members get to know each other as people with a common purpose—to educate all of the children of the community well. Because they are interdependent—what happens in one sphere has some effect on the other spheres—the connections among home, school, and community

Figure 5.2

New Paradigm

Home, School, and Community Relationship—Seamless and Synergistic Collaborative, Interdependent

> Incorporates most transitional characteristics but goes beyond to broaden and deepen relationships and connections

- Focus on total well-being of child, both academic and personal development
- Seamless connections: home, school, and community
- Culture of inquiry, learning, and caring; teachers and parents conduct action research together
- Participatory, personal
- Nonhierarchical, fully inclusive, embracing everyone
- Cultural and social differences valued and preserved
- Power shared with parents and community members
- Parents and community members as partners
- Discovery of common purpose is the goal
- Multiple options, multiple entry points

Everyone together asks: What can all of us together do to educate all children well?

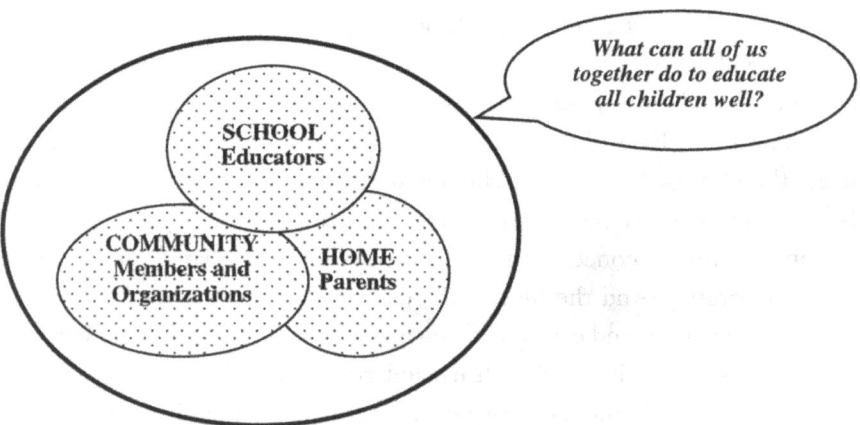

are seamless. People have built the walls that separate one from the other, but, if they honestly answer just one question, they will see why they must tear them down: "What do *our* children need?" The shift in thinking is apparent: not just "my" child, "my" students, "those" kids, but "our" children. Self-interests are not erased, but the common interest becomes an equal priority.

Educational researcher Thomas Sergiovanni explains some of the important characteristics of communities and how they are more than "partnerships":

> In communities, the connection of people to purpose and the connections among people are not based on contracts, but commitments. Communities are socially organized around relationships and the felt interdependencies that nurture them.... Instead of being tied together and tied to purposes by partnering arrangements, this social structure bonds people together in a oneness and binds them to an idea structure. The bonding together of people in special ways and the binding of them to shared values and ideas are the defining characteristics of schools as communities. Communities are defined by their centers of values, sentiments, and beliefs that produce the needed conditions for creating a sense of "we" from "I."[34]

Of course, when groups of people first come together, they do not have the "oneness" that Sergiovanni views as characteristic of communities. Another scholar, J. W. Gardner, suggests finding a common task to begin the process of building community—and that common task is the education of children:

> ... there is no more dependable stimulus for community building than a *common task*—some objective that can only be achieved if diverse elements join in shared action. And there is hardly any common task more deeply rooted in the nation's soul than the future of our children ... the schools are dependably present in every American community and represent the one institution through which all must pass. And finally, creating a sense of community must begin virtually at birth—and after the crucial infant and toddler years, school is a vitally important early experience.[35]

Another bond for people in a synergistic community is the recognition that new learning is essential for developing collaborative ways of working together and for adjusting to the rapid change of living in the twenty-first century. Thus, the community culture in the synergistic paradigm is characterized by inquiry: Everyone—adult or child—is a learner. For many reasons, such as preparation for changing careers or gaining new job skills when factories close, many American adults are already engaged in learning. For instance, according to the National Center for Educational Statistics, 48.1 percent of adults 18 and above in 1999 participated in learning activities during the previous twelve months, 22.2 percent of them for personal reasons.[36]

The learning culture is especially important for another reason: to find more effective ways to educate all children well. While educators have greater professional knowledge about teaching and learning, given the realities of life today and the complexity of human beings, even they do not have all the answers. Students' academic progress is not determined only by what teachers do but is also influenced by personal characteristics, which parents know the most about. And education (in contrast to schooling) happens not only in schools but also in the community. Children are learning everywhere, everyday, from a variety of sources, including their peers, video games, and television and movies.

Since there are so many direct and indirect influences in the complicated world children live in today and since the process of learning is so complex and personal, people have to join together to figure out what children need to become the kind of adults we would like them to be. Therein lies the possibility for synergy because, as educational researchers Bruce Joyce and Emily Calhoun explain: "Synergistic environments—those characterized by rigorous interchange among people—foster inquiry. Environments that separate people depress inquiry."[37]

The synergistic paradigm promises more than the satellitic paradigm ever could because all of us together can do more than any one of us alone. When educators share power and perspectives with parents and community members, they build support and get needed assistance for doing the most important and complex job there is. When the relationships among the three spheres in a child's life are dynamic, seamless, and symbiotic, everyone does what is best for all children. Everyone respects the many differing personal perspectives and backgrounds people in any community may have—and they seek to understand multiple points of view. Everyone recognizes that children's learning depends on affective as well as academic factors and that, while children are in school only some of the time, they are being educated all of the time.

Through conversation, teamwork, and collective action, people work for needed change. And while finding common ground may not be possible, given time, consensus about some ways to achieve a *common purpose* will emerge to serve as a foundation on which to build.

Community Schools Are Nothing New

The new paradigm we propose is really not a new idea—even though to some readers it may appear to be a radical and drastic change from existing

> **Sidebar 5.4**
>
> ## The Principles of Community Schooling
>
> In the social ecology of community life, the community school serves the needs of the entire community by improving and developing it. [Researchers] enumerate the following common principles or elements governing the development and maintenance of the community school:
>
> - Self-determination, in which the members of the local community identify their needs and desires. This is especially important for parents, who now can have a greater voice and take a fuller role in their children's education.
> - Self-help efforts through which people in the community both develop a capacity to solve their own problems and take a greater responsibility for their own well-being.
> - The development of leadership in the community to carry out these self-help and community improvement efforts.
> - The localization of programs and other efforts closest to where people live.
> - The integration of the delivery of services and the maximum use of resources through collaborations of all agencies and organizations within the community to ensure that all needs are met and that all resources are used well to meet local needs.
> - The inclusion of all members of the community to ensure the full development of the entire community.
> - The responsiveness of public institutions with a responsibility to meet the changing needs and desires of the community.
> - The provision of formal and informal lifelong education to all members of the community at all ages.
>
> Source: Erwin Flaxman, "The Promise of Urban Schooling," *Eric Review* 8, no. 2 (Winter 2001): 6–7.

schools. Community schooling takes many forms. In some instances, the school and the community connection came about as a means of using scant resources more effectively or to address a community problem, such as creating a place for "latchkey" kids to go after school. But, in other cases, the community school was established because people in the community believed, as John Dewey did, that the school and world outside should not be as separated as they usually are.

Sometimes these community schools are called by other names, such as "full-service schools," "shared facilities," or "community learning centers," to name just a few.[38] Some are the public schools that all students in the community attend, but others may be charter schools or independent schools. Although these schools attempt to connect schools to the community, they differ in the approaches they take. B. A. Miller, an educational researcher, identifies these three models for rural schools, but they can apply as well to urban neighborhoods: "(1) the school as a community center, a lifelong learning center, and a vehicle for delivering numerous services; (2) the community as curriculum, emphasizing the community in all of its complexities as part of students' learning in the classroom; and (3) the school as a developer of entrepreneurial skills [students learning how to support themselves in the community, for instance]."[39] Timothy Collins, another researcher, notes that a fourth model might be using new technologies to preserve the community while linking students to the rest of the world.[40] He also notes that these models can be interrelated.

Regardless of the model chosen, a community school attempts to serve students and the community. Thus, different communities will make different decisions, depending on their particular needs. For instance, Cambodian and Hispanic families in Lowell, Massachusetts, created their own charter school because they felt their children were not being educated well in the public schools. The school curriculum preserves the cultures by requiring that students study either Khmer or Spanish for several 20-minute sessions each week and by helping students learn also to read and write in English.[41]

On the other hand, a public school, the Geraldine Palmer Elementary School in Pharr, Texas, has an active after-school program:

> [At 4 P.M.] While children at most schools across the United States are leaving school for home, the students at Palmer are engrossed in their after-school activities. In one room, students carefully practice their steps for an upcoming ballet folklorico performance for parents. "Muy bien," says their instructor, a parent who was recruited for her talents as a trained dancer. Another parent nods in agreement while she sits in the corner, carefully stitching the students' costumes.
>
> Down the hall, in a former storage room, students are applying math skills to inventory the canned goods in the school's food pantry. They are preparing for the next day's distribution to neighborhood families in need of food.
>
> In the office, several students huddle around [the] phones ... calling local businesses, soliciting donations for their school uniform drive for needy students. ...[42]

Funded by federal dollars, the extensive opportunities available in Pharr illustrate how community schooling not only improves education for students but also helps to change the lives of some poor families living on the Texas-Mexican border. This school is one of 118 schools that are "Texas Alliance Schools." The Alliance, which assists mostly Latino and African American families in poor communities like Pharr, is based on the concept that "makes the schools community centers and gives parents and community members, alike, a voice in the education of their young people."[43]

Regardless of the impetus for making the connections between school and community or the extent of these connections, people who wouldn't normally encounter one another will do so—and these interactions, casual and intermittent at first, may be the small steps that lead to greater integration of home, school, and community in the future. For example, because state funds for school construction are so limited in Maine, many communities have chosen to raise money locally (through property taxes and private donations) to add auditoriums to new schools that the community will use. The downside to this practice is that wealthier communities end up with more resources than poor communities and so the gap between the haves and the have-nots gets larger. Yet any school can use its facilities—whatever they may be—to serve the community.

One Maine community did raise an additional $7.8 million for its new high school so it could build the kind of facility residents wanted, including extras the state wouldn't fund. Yet the most innovative feature of the new Camden Hills Regional High School was included to make it more inviting to the community, a beautiful cafeteria that is open to the public from 7 A.M. to 8 P.M. So many community activities are scheduled in the school, the principal says, "This building is almost as busy every night as it is every day."[44] Imagine the informal conversations that can happen anywhere just by opening some existing school cafeterias to the community.

Educators in schools that have begun to establish closer ties with their communities see many benefits. One superintendent in Buffalo Grove, Illinois, says: "The more people use our schools, the more they associate with us, identify with us, and use our services and facilities, the better chance we have for enlisting the community's support when we need it." An assistant superintendent in Pontiac, Michigan, notes that "[t]he older community members like to have a place to get together and talk that is safe. The school is a safe place. So they gather at the school and chat and have all sorts of social activities." And a superintendent in Alaska describes his school as "the cultural center for our village—a museum, a library, and a place where our children can learn about their heritage. . . . Community use really helps with increasing support for our facilities. Our curriculum is based on community sharing and involvement."[45]

The community school model has been a tradition in some communities for 100 years; in others it is much younger. Moving closer to the truly seamless, synergistic connections necessary to educate all children well has begun, but there is more to do. And that process for communities still burdened by the old paradigm of home, school, and community relationships begins not with rational plans, committee meetings, reports, consultants, or legislation, but with individuals fully using their imaginations.

IMAGINING MORE IDEAL RELATIONSHIPS AMONG HOME, SCHOOL, AND COMMUNITY

We cannot just tack on discrepant ideas; we cannot add the idea that the earth is round to the idea that it is flat.

—Elizabeth Kamarck Minnich, *Transforming Knowledge*

The analogy Elizabeth Minnich uses to emphasize the need for reconceptualizing taken-for-granted assumptions about what counts as knowledge applies equally well to reimagining the home–school–community relationship. As parents, educators, and community members invent new ways of connecting and put them into practice, they engage in evolutionary change. At some point it may become clear that what we have always thought of as school—or public education—in its present form no longer makes sense. To complete the transformation to the new ways of working to educate all children well (the idea that the earth is round), we will have to give up the traditional ways we "do education" today (the idea that the earth is flat).

What the synergistic model might look like in practice is not clear because it does not fully exist—and the steps a community takes as it moves toward the ideal, as yet unknown, will lead to other steps. Yet these later steps cannot be foreseen because they will grow out of what came before. It's as if someone took a path that led in the direction of a golden building, high on a distant hill. At various points along the way, however, it turns out that there are choices to make: to go one way on what Robert Frost called "the road less traveled" or another that is more clearly marked but seems to veer off what looks like the most direct way to the golden building, still quite far away but visible in the distance. None of these options could be known at the beginning. They would be found only when one had gone some distance from the starting point.

Yet even without a practical example, we can begin to imagine what a new paradigm might mean for children. One way to begin is to think about what children need in order to learn and develop by answering some questions. Readers should mentally add their own ideas to the following responses.

What do children need in order to learn what they need to know to be successful as adults?
Food, shelter, good health, adults who love them to serve as positive role models

What do families need to care for their children well?
Food, shelter, good health—and the knowledge, skills, and resources necessary to provide for children, friendship and support from other adults

What do we want children who have been educated well to be able to do as adults?
Earn a living; participate in civic life; raise their children well; feel confident, competent, and welcomed as members of the communities in which they live; enjoy friendships with other adults, care about the well-being of all children

What special services do some families, individuals, and/or children need?
Special-education services? Psychological counseling? Healthcare? Rape counseling? Rehabilitation programs for substance abuse? Day care for working parents? Parenting education? Job training? Literacy programs? ESL instruction? Legal aid?

What is the community already doing to provide support?
Brainstorming a list of what's already being done in a typical community is surprising—because many individuals, groups, and agencies do serve children and families in a variety of ways. We invite readers to add to the list following.

    Schools and associated groups, such as band boosters and the PTA
    Social service agencies (federal, state, local, public, and private)
    Health services (federal, state, local, public, and private)
    Juvenile justice agencies
    Service clubs, such as Rotary and Kiwanis
    Big Brother/Big Sister
    Habitat for Humanity

Food banks
Homeless shelters
Subsidized housing
Literacy programs
Support groups (ranging from Alcoholics Anonymous to weight-loss groups and families coping with a particular disease)
Head Start, day-care programs
Youth groups, such as Scouts, 4-H, boys' and girls' clubs
YMCA/YWCA
Religious groups
Museums, theater groups, and other cultural organizations

The existence of so many resources seems very promising until one asks: **How connected are all these efforts?**

How easy is it for any individual to know what is available and to gain access to a particular service or program?

How much do people involved with one institution, organization, or program know about the efforts of people doing similar work with another institution, organization, or program?

The problem should be clear. Despite many efforts by many individuals, groups, agencies, and organizations to help solve problems, these services are fragmented and disconnected from each other—and perhaps unknown by the people who need them most.

Questions as Catalysts for Creative Thinking

> *In a time of exploding change—with personal lives being torn apart, the existing social order crumbling, and a fantastic new way of life emerging on the horizon—asking the very largest of questions about our future is not merely a matter of intellectual curiosity. It is a matter of survival.*
>
> —Alvin Toffler, *The Third Wave*

> *... [T]o live the deep questions we must develop a taste for paradox—not least the paradox that some questions have no conventional answers.*
>
> —Parker Palmer, "Evoking the Spiritual in Public Education"

One of Alvin Toffler's "very largest of questions" might be "What kind of society do we want to live in tomorrow?" Thinking about a response to this question begins with recognizing the problem presented in the previous section. Despite all the good work many people are doing, their efforts are disconnected. No one really knows what others are doing to address the same need.

There are many ways to improve the lives of adults and children if we think about making more connections, doing things differently. We invite readers to add to this beginning list of what-ifs:

- What if families who needed assistance could go to one place for help in getting food, jobs, or health services?
- What if new school buildings were planned as community centers? What if all school buildings were used 24/7 all year? Could they become what the town square was 100 years ago?
- What if schools and shopping malls had health and social-service information centers?
- What if the community library and the school library were one and the same? What if all shopping malls had education centers? What if computer terminals connected to the library were located there?
- What if day care and supervision for latchkey kids were available to everyone who needed them?
- What if senior citizens' centers and literacy programs were located in places where children were learning?
- What if every child had an adult mentor? Every family needing assistance had a supporting "friend"? Newcomers from another country had someone committed to helping them adjust to life in a new place?
- What if community members—representative from service clubs, social-service agencies, schools, and other groups—joined forces and *all* worked on the *same* problem for a year?
- What if "school" had expanded hours so that adults and children could learn together?
- What if the school cafeteria were open at night to serve everyone in the community, perhaps before they attended an evening meeting or other event? Would senior citizens feel as isolated? Would parents learn from other parents? Would everyone find new friends?
- What if there were tutors or homework help available to children in the evenings in schools, shopping malls, and libraries?
- What if school guidance counselors had evening hours in some of these places?
- What if every new mother had someone to help her develop good parenting skills and learn about community resources she could tap?

- What if the juvenile justice system focused on education rather than incarceration? Found mentors for every child who gets in trouble?
- What if people joined together to influence advertisers and TV executives to include more positive role models for children in their programming? Less violent video games?
- What if there were more opportunities for adults and children to learn together?

This list is certainly not exhaustive, but it suggests that, if we thought about education more broadly and creatively, we might be able to educate people—both children and adults of all ages—more effectively. And communities wouldn't have to spend more money, but they would spend the funds they have now differently. These resources combined with the volunteer energy available would enable communities to do more without increasing taxes. Given the technology available today, one can easily imagine a computer program that inventoried and described everything everyone is doing. Then someone who needed assistance could type in a need and immediately see the potential resources available for help in meeting it.

Synergy To Create a "Community Home" for All Children

> *Community emerges when we are willing to share the real concerns of our lives. But in our society, you are reluctant to bring your concerns to me because you fear I am going to try to "fix" you—and I am reluctant to receive your concerns because I fear I am going to have to "fix" you. We have no middle ground between invading one another and ignoring one another, and thus we have no community. But by practicing ground rules that release us from our mutual fears, by teaching us how to live our questions with one another rather than answer them, the gift of community emerges among us—a gift of transformation.*
>
> —Parker Palmer, "Evoking the Spiritual in Public Education"

Synergy requires that everyone work together, but also, as the what-if questions suggest, *education*, not just *schooling*, must guide our thinking. We need to think differently about schools and more broadly about education for very practical reasons. Because resources are limited, we need to spend the dollars we have to get the most for our money.

Even more important, though, we need to remember that the future we all will have is in the hands of the children we are raising today. The future we get will be the results of our action—or inaction—to help all of our children become the best they can be. This important challenge requires asking what Alvin Toffler might consider the "the largest of questions." What is a democratic society? What kind of world do we want to live in? What do our children need? What would it mean if we created a "community home" for children? How can we collaboratively work to serve all children well?

Where do we start on this journey toward the synergistic ideal? Before we can begin to reap the benefits of such a relationship among home, school, and community and parents, educators, and others working for a common purpose, we first have to build a foundation of trust and respect. This topic is the focus of chapter 6. The following chapters profile two schools that have already done a great deal to open their arms to people outside.

6

GETTING STARTED: BUILDING A FOUNDATION OF TRUST AND RESPECT

I must trust like I've never trusted before.

—Parent of a kindergarten child prior to September 11, 2001

Millions of parents throughout the country and around the world send their children off to school each day trusting that they will, at a minimum, be safe. It is no wonder that parents consistently rate their schools higher than they rate schools in general.[1] For how could they live with themselves if they knowingly sent their children to schools they did not trust? Spoken after the school shootings at Columbine High School, the parent's words just quoted have added significance after the unspeakable horror of September 11, 2001. With our traditional American sense of security crumbling, parents have to trust as they never have before that the schools they send their children to will, above all else, keep their children safe and secure.

Many parents were reassured when they heard how compassionately and capably many of their children's schools—teachers, principals, and others—dealt with the situation. Teachers in lower Manhattan led their students to safety amid the chaos that surrounded them. Plans were put into action to care for children whose parents might be affected by the situation. Teachers throughout the rest of the country assured their charges that they were safe. Teachers, emotionally shaken, put the needs and well-being of their students first. Principals relieved teachers who were having a hard time coping. Teachers were given guidance about how to respond to children's fears and questions. Counseling for individual children was made available. Letters were sent home to parents with suggestions about how to help their children cope. Children and families were invited to participate in school memorials and to provide material assistance to the volunteers in New York City.

But there were other stories. One school ordered teachers not to talk about the situation with the children, period. It didn't seek advice from psychological support staff about how to help students cope. It didn't send information to parents about how to help their children deal with their fears. And it did not make any special efforts to help students feel that they could contribute in this time of great national emergency. One angry parent reported that the school her child attended did not offer to make counseling available to children who were troubled by these events. This parent was angered by this lack of leadership at a time when her child needed greatness.

These words are being written only a week after the terrible tragedy of September 11th, so emotions are volatile—as is the situation. While our leaders speak forcefully about the need to respect all of our citizens, American citizens of Middle Eastern backgrounds and Muslims are being attacked physically and verbally. Even Americans from India are being attacked because some of us don't recognize that they are most likely not Muslims or Arabs. Muslim parents are afraid to send their children to school because they might be harassed. These circumstances underscore the importance of this chapter. Our schools must be centers of multicultural understanding and sensitivity. We must help our children and their families—and ourselves—to learn about the different people who are represented in our communities. All members of our community must feel safe. The challenge for us is: How can we build an inclusive school community grounded by trust and respect that cannot be shaken even in the most trying of times?

WHY IS A FOUNDATION OF TRUST AND RESPECT IMPORTANT?

Everything we know about school-community, professional-nonprofessional relationships (in the past and now) permits the predictions of problems, among which the absence of trust and respect is the most troublesome.

—Seymour Sarason, *Parental Involvement and the Political Principle: Why the Existing Governance Structure of Schools Should Be Abolished*

While we believe that all of us—teachers, parents, and community members—are responsible for educating our children, this chapter zooms in on the relationship between educators and parents. This relationship is central to the project of educating our children. Without strong, trusting relationships between parents and educators, nothing else matters. Community connections, although important, can't compensate for poor relationships between families and schools.

Researchers, politicians, teachers, and parents all agree that parents and teachers must work together as partners in the education of children. Central to such a partnership is a strong working relationship among teachers, principals, parents, and students. Trust and mutual respect are the very foundation of this relationship. Without these qualities open communication, which is essential, is not possible. A lack of trust and respect fosters questions about motives and misunderstandings rather than the open and honest communication necessary to address the issues. Researchers who study schools as organizations have identified trust as a key characteristic of successful schools.[2] Studies of teacher and parent expectations of each other also demonstrate that both parents and teachers want to be respected and trusted.[3]

What exactly do we mean by *trust* and *respect?* The fact that there are many definitions of *trust*, suggests its many dimensions. Organizational theorist A. K. Mishra, for instance, captures this multidimensionality: "Trust is one party's willingness to be vulnerable to another party based on the belief that the latter party is (a) competent, (b) reliable, (c) open, and (d) concerned."[4] This notion of vulnerability seems to underlie most definitions of trust and probably is what makes trust so difficult to establish between teachers and parents.

Social psychologist John K. Rampel and his associates suggest that there are different levels of trust: predictability, dependability, and faith.[5] Predictability relies on evidence of concrete observable behaviors. A parent thinks, "I trust you because you listened to my concern yesterday and modified my child's homework assignment." Or a teacher thinks, "I trust you because you sent in the items that I requested for today's lesson."

Dependability moves from specific behaviors to a belief that the other person is trustworthy. Building trust does not happen in an instant; it takes time. After a series of incidents or interactions where the other person behaves in a trustworthy manner, the parent or teacher gains confidence in the other's trustworthiness. For instance, a teacher might note that the parent always makes sure her son does his homework. The parent, on the other hand, observes that her son's teacher always calls her if he is having trouble with an assignment.

Faith, the third component, goes beyond a consistent pattern of observed behaviors to a general sense of trust. Both parents and teachers come to feel "I trust you because of who you are." Incidents in which trust might be tested are not viewed as violations of trust because parents have confidence that the teacher's motives are honorable: "Even though Melissa came home and complained that you didn't help her with a math problem she didn't understand, I know it was an oversight. I still trust you to do what's right by my child." And the teacher assumes the trusted parent has a good reason when something isn't done: "Even though your child came in unprepared today, I realize that last night was probably a busy night for you. I still trust you to carry out your responsibilities."

Trust can also be viewed on a continuum of low- to high-trust relationships. For instance, a low- to medium-trust level would look something like this. Parents and teachers view each other from a perspective of suspicion and look for proof that "I can trust you." Every action or interaction is watched closely for indications that a parent or a teacher is trustworthy. So when a teacher, focusing on getting the morning started, responds too abruptly to a parent who stops by at the beginning of the day to ask a question, the parent thinks, "See, this teacher doesn't pay attention to my concerns." Or when a parent juggling the different schedules of her three children, working late, and caring for a sick mother forgets to send in her son's field trip permission form, the teacher thinks, "I knew it. This parent doesn't care." As one teacher recalls, "I would be questioned a lot as to what my motives were; why I didn't push more. [They were] not allowing me to teach and guide the child and that class as I saw best, and they would question my judgment on their child's behavior."[6]

Using Rempel's framework, Kimberly Sue Adams, in her doctoral dissertation completed at the University of Minnesota, describes the problems created when parents and educators have few opportunities to interact:

> Due to limited amounts of contact in the parent-teacher relationship, trust may often remain stalled at the lowest level, predictability. When this happens, parents and teachers continually search for behaviors as evidence of the other person being trustworthy. This failure to move on to the realm of personality attributions makes trust in this relationship extremely vulnerable to fluctuations in behavior by either the teacher or parent. One false step

can move a potentially trusting relationship in the wrong direction. Holmes [and] Rempel... warn that once there is a serious breach of the trust relationship, the relationship is at serious risk of failing.[7]

On the other hand, a high-trust relationship starts with the assumption that the other person is acting in "my" best interests. Therefore, when things don't go as a parent had hoped, she doesn't blame the teacher. For instance, if a child comes home and tells the parent that the teacher punished her for something she did not do, the trusting parent, rather than approaching the teacher defensively and angrily, as one in a less-trusting relationship would do, wants to find out more about what happened and makes an appointment to talk with the teacher. And the teacher in this high-trust relationship would welcome the opportunity to talk about the situation with the parent, to find out more about the child's perspective, and to come to some common understanding with the parent about what happened and what should be done about it. Now, it may be that the parent will find out that her child had indeed done what the teacher thought she had, or maybe the teacher will find out that there were extenuating circumstances that she didn't know about. In either case, both teacher and parent would leave the meeting believing that they had each learned something from the other about the situation and feel good about the outcome.

Mutual respect and trust are related phenomena, but determining which comes first is difficult. Perhaps they are reciprocal; each contributing to the other: "If I respect you as a person and as a professional, then I can trust you. If I trust you, then I must respect you." It used to be—or so we seem to believe—that parents automatically respected teachers because of their position in society. And that trust was derived from this respect. "I respect you as a teacher, and, therefore, I trust that you will do what is best for my child." However, that sense of respect has diminished. Since the 1960s, when the public began to question all of our institutions and professionals, the traditional respect given teachers has decreased. Very often today we hear teachers lament the loss of this respect. As a result parents, rather than automatically trusting that the "teacher knows best," begin their relationships with teachers with a low level of trust. They look for evidence that teachers are treating their children well and that they have the knowledge and ability to help their children learn. Teachers also used to respect the capability of parents to raise their children. When communities were more homogeneous, and teachers more closely resembled the children they taught, they were more likely to trust parents.

Today, however, the situation is different. Society is more diverse. With the changing face of our population, teachers are increasingly teaching children from backgrounds different from their own. They are concerned that children are not being raised with the values teachers hold dear and that parents too often blame the schools for their children's failings. Thus, teachers, too, now look for evidence that parents are trustworthy; that they support their professional decisions, encourage children to do their homework, and make sure that children are well cared for and prepared to come to school each day with adequate nourishment and sleep.

WHAT TRUST AND RESPECT MEAN TO PARENTS

. . . with teachers dealing with so many kids at once, their schedules disrupted by special programs and tests, having to teach not just reading and math, but where to put the knife and where to put the fork, I don't feel they can know my child as well as I do. So how can I really trust them?

—Beth Wolaver, Parent, quoted in
Linda Perstein, "Suspicious Minds"

If we're a parent who's really in tune with our child, and sits with him every single night and helps him do homework and sees where his strengths and passions are, I think we know that child best.

—Elena Valesco, Parent, quoted in Kathy Murfitt,
"Addressing the State of Parent-Teacher Relations"

Parents want to trust teachers to keep their children safe and secure, to know their child intimately—almost as well as the parent does. They want to trust teachers to plan educational programs to meet their child's particular needs. They want teachers to listen to them with respect, without becoming defensive, when they try to share what they know about their child. They want to believe that their children are entrusted to competent and caring teachers. They want to be respected as co-equals in the education of their children.

Parents want to believe that their children's teachers, most of all, will keep their children safe. For instance, when parents at Greenbrook Elementary

School, profiled in chapter 7, watched their five-year-olds first walk down the very long corridor to their kindergarten classrooms, they wondered how they would learn to navigate the strange hallways and byways of the school. Would they get lost? Would they be frightened? Surely most kindergarteners learn very quickly how to do this. But a parent's first thought is for her child's safety. After the Columbine tragedy, parents' concerns for the safety of their children escalated. As one parent said, "Who would have thought that I'd have to worry about sending my child to school?"[8]

Parents want to be assured that in the process of teaching an entire class, the teacher also gets to know her child well as an individual—what he likes to read, what she does at home for fun, what subjects she finds hard, who her friends are, what his family did over the summer, how she learns best—and then uses this information to tailor a learning program to meet his specific needs. One parent, who knows what motivates her son, wants the teachers to know this, too: "My son loves motorcycles. Absolutely loves them. A mechanics class where they have them rip down a motor and see how it works to see a two-stroke engine versus a four-stroke engine versus an eight-cylinder car—things like that would be fun and they'd learn something."[9]

Since parents are not in school all day, they get their information about what happens there from their children—and what they hear may cause them to lose confidence in teachers. For example, when a child comes home from school day after day complaining that her middle-school teachers don't know her name, that they only know her from the seating chart in front of them, the parent believes that these teachers don't care very much about her child or know her well enough to provide a program to meet her individual needs. Of course, teachers who see over 100 children a day can't know each child and each family well. School bureaucracies interested in efficiency have chosen to build bigger and bigger schools, believing in the financial benefit of scale. Some school districts, however, such as Crossroads School, profiled in chapter 8, have begun to reorganize large schools into smaller learning communities so that teachers can get to know each child well.

Parents want to be listened to respectfully—and not defensively—when they come to teachers with information about their child. They don't want to be dismissed as pushy parents. When that happens, the chances of building a trusting relationship are slim. Teachers also need to ask questions, however, not just assume they know what the parent means. For instance, one parent said to a teacher, "My child is very good at math. I hope you will challenge his ability." Rather than exploring what that parent meant *with* the parent and thinking through an appropriate program *with* the parent, the teacher (perhaps thinking "I am the expert here") responded defensively with an explanation of her math program for all children. The parent left wondering if the teacher really listened to what she said. Parents have learned that they have to tread very lightly when they try to make educational suggestions to teachers, since many teachers are quick to think that their educational expertise is being challenged. It's hard to trust someone who doesn't take your ideas and suggestions seriously.

Parents want to trust the teacher to treat their children fairly. One of the biggest areas of stress in relationships between parents and educators is the "she said . . . he said . . ." scenario. Teachers and administrators complain that parents are too quick to believe their child when they are accused of doing something wrong. Because they don't believe that teachers know their children as well as they do, parents, on the other hand, tend to believe their child. Parent Beth Wolaver says, "If the teacher and [my] child have a history of 'not clicking,' I might side with my child. If there's no bad-clicking history, I'll check it out. But my instinct would be the child is right."[10]

Parents want assurance that their children are in the hands of competent and caring teachers. And they believe it is their responsibility to advocate for their children if they believe teachers are not. The dirty little secret of schools is that there are some teachers—surely a minority, but still some—who should not be in the classroom. Many middle-class and affluent parents know how to be proactive and make sure that their children don't get these teachers by approaching the principal with their request. The children whose parents don't have the know-how to negotiate the school's bureaucracy get stuck with the ineffective or incompetent teachers.

Many parents keep their eyes open for signs that something is wrong, but their interpretation is limited by their own knowledge and experience. Although well-educated parents may know more than those who didn't finish high school, they may evaluate what teachers do based on models of schools and classrooms they attended in the past, which can be quite different from what's happening today. They look first at how their child is doing. Is he or she learning how to read? To sound out words? To write? To spell correctly? To calculate? To do the things they, the parents, think are signs that their child is being properly educated? If they don't see what they expect, they begin to question the teacher's competence, to question the professional decisions the teacher makes. Do the curriculum and teaching strategies match the parent's mental models of what good teachers do? If the school's curriculum and textbooks include material that parents don't believe is accurate or relevant for their child, they become suspicious. For instance, when one parent learned that the school was teaching students that Christopher Columbus did NOT discover America, he became furious—and suspicious of the school's motives. He began to question the school's ability to teach his children American history from the same perspective from which he had learned it.

Teaching strategies also can be problematic for some parents. When children are asked to write essays in math class, when reading is not taught strictly through phonetic analysis, or when spelling isn't stressed in a child's first draft of a written assignment, parents who had been taught in these ways question the teacher's competence. Many times, when a school is implementing new teaching methods for the first time and teachers are learning how to use these new strategies, parents become concerned about their children being used as "guinea pigs." They need reassurance that their children will not suffer as teachers learn to use these new strategies. As one parent put it: "We can't hit the rewind button and do the year all over." When teachers can't clearly articulate their rationale for doing what they are doing in language that parents understand, the possibility of building a trusting and respectful relationship between parents and teachers is threatened.

As has often been said, parents are a child's first teachers. When the child enters school, parents don't disappear from the picture as the child's teachers. Rather they have to make room for additional adults in their children's lives. As one parent said, "My child needs other strong adults to help guide her." When parents invite teachers to be co-equals with them in the education of their children, they do not abdicate their role as "first teachers" when their children walk into the schoolhouse door.

Unfortunately, many times teachers invoke their professional status when talking with parents and give parents the impression that—rather than seeing this as a co-equal relationship—they (the parents) have nothing to contribute to the conversation. As one parent said, "There were instances from both my daughter's and son's experiences where the teacher said this is the way things are and you don't understand. The curriculum [was] drawn up by professionals. You don't know as much as we do."[11]

Parents want teachers to recognize their knowledge and to respect them as partners in the education of their children.

Sidebar 6.1

Minority Parents Take the Initiative

When parents of freshmen at Berkeley (California) High School realized that their children were failing, they decided to take action. Near the end of the first semester, half of the school's 300 African American ninth graders were not passing English, math, and/or history. The situation called for action because this class would also be the first one to face California's new exit exam in 2004. Parents joined together in a group they called Parents of Children of African Descent (PCAD) and went to the principal to discuss their concerns.

Challenged by the principal to develop a plan, PCAD did just that. On Martin Luther King Day, PCAD convened a luncheon meeting of parents and community leaders to present the PCAD Intervention Plan, which the group wanted in place at the beginning of the second semester just two weeks away. The program, designed for ninth graders who were failing two or more subjects, would create "an alternative learning community" inside Berkeley High. Students accepted for the program would be in small classes taught by teachers chosen for their commitment to improving the achievement of at-risk students and assisted by student mentors. So that students would be able to catch up with their peers and be ready for tenth grade in the fall, the program would extend into the summer.

Katrina Scott-George, a parent and later a teacher in the program, explained how crucial adults were to the program's success.

> We want to create parent demand for education.... We want to build a network of parents reaching out to parents—supporting their kids' education, making the school respond to their kids' needs." Toward that end, each student in the PCAD program would be assigned an adult "learning partner" recruited from his or her family, school, or community. Parents and guardians would sign contracts requiring them to respond promptly to teachers' calls home. A "We Care" campaign would be launched to counter what PCAD calls "the subculture among African American and Latino students that often penalizes success.

Since limited school funds were available, PCAD raised the rest of the money needed to begin the program from other public and other private sources. Five new teachers were hired, and 50 students were recruited for the program, which students later named "Rebound." Each teacher was responsible for 10 students, but the focus was not just on improving their academic skills but also on working closely with parents and community agencies and so that they could help students with any aspects of their lives that might be barriers to school success. Parents (or guardians) were required to be involved. In fact, because many of them accompanied children on the first day of the new semester, it was the first time students had come to school on time all year. Writing about the program, freelance writer Meredith Maran, who spent the year writing her book, *Class Dismissed*, at Berkeley High School, notes that "[t]eachers [would] spend their days struggling to engage kids who've never been taught successfully,

(continues)

> **Sidebar 6.1** *(continued)*

their nights and weekends building relationships with families who've had reasons not to trust them."

The program plans sometimes changed as teachers worked with these needy students, but the teachers created a caring environment in which many of the students began to improve. At the end of the school year, according to Scott-George, about 25 percent of the students "have been significantly turned around.... They have the tools that will allow them to be successful: emotional ability, study skills, and basic skills in math and English. Another group, about two-thirds of the kids have made some kind of commitment to their own education. They may not have the skills, but we can see huge changes in them. More than half . . . are going to be on track for graduation by the end of the summer."

Rebound existed for only one year because Scott-George says:

> That was the original plan. The PCAD Intervention Plan is not about a single program like Rebound. We were working to try and change how students are educated, to provide a model for student-family-school collaboration. We provided the model. What the district will do with what we've given them remains to be seen.... Change on a bigger scale can come about through creating parent demand for education. That's why our plan next year is to build on this group of parents we've established strong ties with.

Recognizing the value of the program, the principal announced that the school would design another program to help students who needed help in the core curriculum. Because of limited funds, however, the school plans a scaled-down version of Rebound called "Tracker" in the fall using teachers already in the system to staff it.

Programs like Rebound, though, are important not only for minority students but also for their parents. As Scott-George explains: "There's such a history of distrust of Berkeley High by African American parents.... They haven't felt the school is interested in educating their children, so it's been hard for them to present school positively to their children. Rebound is causing parents of color to believe that with their own advocacy of their own children, a school can work on their behalf. Longterm, that's where our greatest hope lies."

Source: Meredith Maran,"Damage Control," *Teacher Magazine* (August/September 2001): 25–37. Meredith Maran is the author of *Class Dismissed: A Year in the Life of an American High School*, published by St. Martin's Press.

WHAT TRUST AND RESPECT MEAN TO EDUCATORS

I loved the kids. It was the disrespectful way some of the parents treated me that was the problem.

—Patricia Woolsey, former teacher, quoted in Murfitt, "Addressing the State of Parent-Teacher Relations"

Parents should trust educators to make the right decisions because they are the professionals.

—High-school teacher, quoted in Jean L. Konzal, *Our Changing Town, Our Changing School*

Educators want parents to treat them with respect. They want parents to acknowledge their professional expertise and trust that they will do what is right for individual children even as they teach all the children in their class. They want to be able to trust that parents will support them. They want to be able to trust that parents will do their jobs well—to raise children who respect their teachers, to make sure children do their homework, to be prepared—physically, mentally, and emotionally—when they come to school.

Too many teachers and administrators have been burned by the sharp tongues of angry parents. These experiences leave them emotionally drained and on guard to protect themselves against future attacks. Even though only a small minority of parents resorts to verbal assaults, teachers experience such emotional vulnerability from them that they begin to feel suspicious of all parents. "Who will be the next parent to verbally lambaste me?" they wonder. One teacher said, "It takes such an emotional toll on me to respond calmly and professionally when I am verbally attacked."

Parents at both ends of the socioeconomic continuum approach teachers inappropriately. Parents from working-class backgrounds, feeling disempowered in their relationships with educators, may resort to angry outbursts. One mother, for instance, explained:

> I grew up on welfare. My family was on welfare. We had no money. There were five of us who shared a bedroom and we were treated like dirt in the schools. I always felt that because I was on welfare I didn't get the respect that everybody else got. Nobody is gonna treat my kid like dirt. I want respect for myself and I want respect for my kid. I went in there on Friday to

talk with them about a problem my kid was having and I left in tears. I was crying because I felt so stupid because they made me feel like I was nothing.[12]

On the other hand, professional parents, used to dealing with others from a power position, have no qualms about questioning teachers. Teachers in affluent communities, according to veteran teacher Linda Lieberman, "are seeing a bigger push on academics, on questioning how testing is being done, questioning how children are learning, questioning the background of teachers. When I started, parents would never question what the teacher did. [Now] advertisements and magazine articles are constantly saying you need to be your child's advocate."[13] And some do not hesitate to use threats of lawsuits to get their way. One teacher from a school district that serves well-educated, professional families said: "We deal with parents who are power brokers in their work world, and they tend to bring their power to bear in their child's classroom, where power doesn't have a place. Some parents think that . . . threatening someone with their lawyer . . . will ensure better service for their child. But what happens is nobody wants that child in their classroom."[14]

When teachers feel that the professional parents in the community don't respect them, relationships become strained and interactions, tense. A recent conversation with a retired superintendent of an upper-middle-class suburban school district sheds some light on some of the dynamics that impact teacher/parent relationships in such communities. Parents in her community—mostly businesspeople, doctors, and lawyers—assertively approached teachers when they had concerns. Well educated, articulate, and informed, they did their homework on the issues and had personal and professional contacts in the world of education. One parent, for instance, quoted her uncle, a professor at the University of Chicago, during one interchange with a teacher. These parents expected well-thought-out responses to their questions. When teachers, caught off balance, were unable to clearly articulate their rationale for what they were doing in the classroom, they lost credibility in the eyes of these parents.

Even when teachers represent their profession with competence and poise, however, parents from the community's power elite may still look down on them. For instance, one board member—an attorney—actually swiveled his chair around and turned his back on a teacher (who many years later became the superintendent of this district) as she tried, during a board meeting, to explain the report she had prepared for their consideration.

These "power plays" may exacerbate teachers' inability to be articulate, thus increasing parents' perceptions of their incompetence.

Educators want their professionalism respected. Over the past 15 years teachers have been encouraged to become master teachers—to be the experts about what they teach and how they teach. Many have taken this challenge seriously and joined study groups; become part of local, regional, and national networks of educators seeking better ways to teach all children; and tried out new ways of teaching in their classrooms. Many see themselves as researchers of their own practice. Unfortunately, this work has been done behind the scenes, outside parents' sight.

As teachers begin to redefine what good teaching is, parents are left with images of teaching from their days as students. Reading is sounding out words. Writing is good spelling. Math is adding, subtracting, multiplying, dividing. Children learn best when grouped with children of similar abilities. Group work is cheating. Teachers want parents to abandon these old images of good teaching and accept new ones, based on their professional expertise. They want to be trusted.

Parents, however, question what teachers do as a result of what they hear from their children, problems they may see their children having in doing their schoolwork, or observations of poorly implemented programs. Although teachers could use the parent's questions as an opportunity to engage in dialogue, some teachers assert their professional prerogative.

Teachers do not want parents to question their motives. In one high school, parents revolted when teachers redesigned the school's math curriculum. Parents had concerns about many of the new practices: mixed-ability grouping; renaming the courses from Algebra, Geometry, and so on, to Math 1, Math 2, and the like; and the emphasis on group work in the classrooms. What troubled some teachers the most, however, was that their motives were questioned. A teacher said, "It is often suspected that for some reason teachers have ulterior motives for wanting to make changes. The public has a hard time believing that we are looking out for the best interest of their children." Another commented, "An ideal relationship between parents and teachers has to be based on some mutual respect, mutual trust—that we understand their point of view. There's got to be that underlying trust that we're trying to do the best job we can do."[15]

Teachers want parents to teach their children to treat teachers with respect. They worry about parents bad-mouthing teachers at home. If a parent doesn't speak respectfully of a teacher at home, they wonder how the children will respect the teacher at school. If the parent takes a child's side in a dispute with the teacher or questions the teacher's professional decisions, the message children get is that the teacher is neither capable nor worthy of respect. Studies have shown that when parents question a teacher's approach, student achievement may suffer.[16]

Teachers want parents to take good care of their children. Do children come to school ready to learn? Are they hungry? Tired? Fearful? Insecure? If they are, it is hard for them to concentrate on learning. With our increased awareness of child abuse, sexual abuse, neglect, and dangerous situations many children confront every day, many teachers wonder why parents are not held more accountable for their children. Recently New Haven schools have adopted an accountability plan that holds parents as well as teachers accountable for children's educational development. Although the idea certainly is appealing, will rewards offered, such as discount coupons to various stores, really make any difference? Wouldn't opening up a dialogue with parents and providing parent support services to needy parents be far better? Somehow, it doesn't seem likely that casting judgment on parents will result in more trusting relationships or more accountable behavior.

Many times parents who are poor or whose cultures are different from teachers' demonstrate their care for their children in ways that are unfamiliar to the teachers. Too often teachers are suspicious of parents from low socioeconomic communities, and this has dire consequences for children. A recent study showed that when teachers do not trust parents or students who are poor, students' academic performance may be affected. The study asked teachers in a large urban school district to respond to survey items, such as, "Teachers can count on the parents in this school; Teachers think most of the parents do a good job; Teachers in this school trust the parents to support them; Parents in this school are reliable in their commitments; and Teachers can believe what parents tell them." The study found that children's achievement suffered in the schools with low levels of teacher trust in parents and students.[17]

School administrators are often put in tenuous positions—caught smack in between parent concerns and teacher concerns. Teachers complain about principals who don't protect them from parent complaints or who seem to give in too easily to parent demands. One teacher said, "Many times the administration is not behind you."[18] On the other hand, parents also complain about administrators. Going to the administration with concerns about teachers is sometimes a frustrating experience because parents see administrators circling the wagons, supporting their teachers no matter what the details of a particular situation.

School reform efforts may bring about positive change in these relationships. A retired superintendent, for instance, noted that an unintended outcome of school reform and the professionalization of teaching has meant changes in both the principals' and teachers' roles. In the past the principal acted as a gatekeeper, keeping parents out of the way of teachers. Now, however, teachers themselves are expected to interact with parents, to listen to parent concerns, and to explain what they are doing in the classroom and why. Teachers who are unable to clearly articulate what they are doing, of course—especially in upper-middle-class communities—are in danger of losing the trust of parents. But teachers who are able to communicate effectively have a much greater chance of building trust with parents.

PARENTS, TEACHERS, AND ADMINISTRATORS: TROUBLED AND TROUBLING RELATIONSHIPS

Distrust between educators and parents can be found in all types of school communities—in schools that serve children living in poverty as well as in schools that serve children of privilege. In interviews with teachers who taught in a cross section of schools and communities, teachers were asked, "What is the hardest part of your job?" A majority of the responses to this question—from teachers in schools in poor urban and rural communities to teachers in schools in upper-middle-class communities—related to dealing with "difficult" parents. For example a representative teacher response to this question was: "Dealing with parents who refuse to see their children's difficulties whether academic or social. These are the parents who always say 'not my child.'"[19]

In any teacher's room one could probably hear teachers complaining about all types of parents. In both urban and rural schools serving children of poverty, teachers might say: "These parents don't care about their children. If only they would . . ." followed by a litany of complaints, such as "feed them breakfast," "make them do their homework," "discipline them," "have higher aspirations for them." In suburban schools serving children of privilege, some might say: "These parents push their children too hard. If only they would . . ." followed by a litany of complaints, such as "recognize that their children aren't geniuses," "stop telling teachers what to do," "respect my professional knowledge." The issues may be different, but the complaints in both instances usually start with "If only parents would . . ."

Although trust between parents and educators is hard to find beyond the first level of looking for evidence that "I can trust you," both want to trust the other. What are the barriers that prevent this trust from developing beyond the first stage in most schools? As chapter 2 showed, parents and educators see the world of schools through different eyes. Parents are most concerned with the needs of their individual children ("my" child). Educators, on the other hand, have to consider the needs of all of their students, while at the same time thinking of individual children. Educators see the world through their professional roles; parents, through their parent roles. Because their lenses are different, the relationships are difficult. Other factors that contribute to this lack of trust among educators and parents include differences in the quality of school leadership and the lack of opportunities—both formal and informal—for parents and educators to get to know each other well.

SCHOOL LEADERSHIP: PRINCIPALS AND TEACHERS BOTH HAVE ROLES TO PLAY

My job is to unstop toilets, put on Band-Aids, mediate, praise, support, understand, find resources, provide consistency, be cheerful, make sure decisions get made, and, most important, be trusted to hear all sides and be fair. And all of that is based on knowing well—and hence caring deeply about—everyone: students, staff, parents.

—Ann Wiener, Principal, Crossroads Middle School, New York City, "Participatory Leadership"

As the portraits of Greenbrook Elementary School and Crossroads Middle School in the next chapters will show, the principal's role is paramount in creating an inviting school community. Effective school leaders know that a principal must be the master weaver of a complex and textured tapestry of school relationships. These principals lead by example and use their influence to change the behaviors and beliefs of others.

Thea Dahlberg, a teacher at Greenbrook Elementary School, credits her principal with helping her to structure her parent conferences so that she attends to parent perspectives: "Everything I do at conferences I learned from my principal, Patricia Holliday." When principals don't live up to these high standards for leadership, trusting relationships among parents and teachers may falter. These fault lines become especially apparent when educators attempt to change their teaching and learning practices.

Effective leaders help forge an inclusive educational vision. Such a vision is, as organizational theorist Peter Senge says, like a prism, because it reflects the visions of the individual members of the organization.[20] They know how to interpret that vision into day-to-day practice. They understand that they are most powerful when they give away power to others—teachers, parents, and students. They have "people skills," a critical necessity in an organization as complex and people intensive as schools. They know how to encourage, support, and challenge people to do their best. They know how to mediate conflicts so that everyone comes out a winner. They know how to manage change.

Principals who do not have this knowledge, these skills, and these attitudes are not likely to create a climate of trust and respect in their schools. Most principals are under great pressures today: They must make sure their schools meet the new standards set by the state and federal government, that they are safe and secure in times of uncertainty, while simultaneously dealing with the daily emergencies that happen all the time. It's not unlike the circus performer who tries to keep all of the dishes spinning at the same time. Thus, principals may not have the time to develop all the leadership skills they really need. If the pressures of the "dailiness" of schools prevent principals from developing what they need to lead well, many schools will continue treading water.

A principal's leadership will be tested whenever new practices are introduced. As noted earlier, when schools make changes that go against parents' ideas of "good schools and classrooms," parents can get uneasy. For instance, when schools change from tracking to mixed-ability grouping, when they implement block scheduling (longer and fewer classes each day), when they make changes in standards and grading, or when they introduce new curricula and textbooks, some parents will wonder if these changes are in the best interests of their children. Conflict is a certainty if these new practices press against the boundaries of the community's zone of tolerance—the zone of acceptable practices based on the community's unarticulated beliefs about what makes for good schools.[21]

Most parents, for instance, have been educated to believe sorting students by ability is good educational practice. Both parents of gifted children and of children with learning problems believe that ability grouping is the best way to teach students. So when new research raises questions about this practice, as it did in the late 1980s and 1990s, and schools begin to eliminate tracking and ability grouping, parents feel threatened and express their anger.[22] As the story of Springfield Regional High School in chapter 4 shows, parents objected when classes in the school were not tracked. These parents also disagreed with the school's standards-based grading approach because it did not mesh with their unarticulated beliefs about what should be done in good schools. Parents make judgments about the acceptability of new practices based on their mental models of good schools and on their perception of how a particular practice will affect their own children. For instance, Tim Zukas, a parent in another school writing to educators, unequivocally reminds them of his priorities:

> I have a very narrow set of interests. I do not care about the latest advances in brain research. I won't get excited about claims of potential big performance gains. If you tell me about self-esteem one more time, I will become ill. Although I want what is best for all students, I am much more interested in how your proposal will affect my child. It is not that I am uncaring but that I have a special responsibility for her. I will not let you forget her needs so that you can help someone else. If you want my support, talk to me specifically about your reform's impact on her. After you address her needs, I can think about the potential benefits to others.[23]

The approach principals take in making changes can make the difference between parental opposition or support for the innovation. In order to implement any change, principals must first recognize that the practice might cause concern for some parents—that it will bump against the community's zone of tolerance. Principals must recognize the need to build commitment to the innovation among teachers, parents, and students. They need to design a process that invites all of these constituencies to the table to learn about the new practice, to offer suggestions for revisions, and to help implement it. And, finally, they need to plan for an ongoing assessment of the innovation in order to catch and fix problems before they put any students at a disadvantage.

Because of different approaches, the same innovation played out differently in two communities. When one high school in a mid-Atlantic state decided to go to block scheduling, the result was a series of angry community meetings that led to the school board's decision to rescind this policy. On the other hand, in a neighboring community, parents supported the change to block scheduling in part because the principal involved students and staff in a year-long investigation into the costs and benefits of block scheduling and then held community forums presenting what they found. Students were instrumental in explaining to parents what they learned about the benefits of block scheduling. This principal understood how to involve all segments of his community in learning about and making decisions about the new schedule. He also understood the mediating influence of students.

Students play an important role in school change efforts. When students are unhappy or don't understand what they are doing or why they are doing it, they tell their parents. Parents learn a lot about what goes on in school from their children—sometimes getting a big dose of misinformation. When students are involved in planning and innovations are well implemented, parents are more likely to get a positive spin from their children. Too often, schools forget the important role students play in communicating what goes on in school to parents, and principals forget to include them in the process of adopting new practices and programs.

Teachers also affect how parents and students respond to change. The reform efforts of the past 15 years have resulted in teachers taking leadership roles. As teachers take on more responsibility for developing curriculum and co-leading reform efforts in their schools, they find themselves more responsible for explaining new programs and practices to parents. Those who can clearly articulate what they are doing and why they are doing it in language that parents can understand have a better chance of building and keeping trust with parents.

Teachers who can't or who won't clearly explain to a parent what they are doing, why, and especially how a specific practice will help children learn contribute to tensions between parents and schools. For example, many teachers who jumped on the whole-language approach to teaching reading and writing didn't fully understand the theoretical underpinnings of the approach themselves. Thus, they were caught off guard when parents asked for an explanation and rationale for the approach. When teachers lack knowledge, confidence, or effective communication skills, parents lose confidence in them, and confidence is a primary ingredient for trusting relationships.

In some instances, teachers won't explain what they do because they think parents should trust them and not question their professional decisions. But this stance can backfire and lead to parental outrage and revolt. For example, in one elementary school in a northeastern state, teachers implementing a program to teach children the virtue of sharing required all students in the class to put their crayons, pencils, and other materials in a common box for everyone to use. Parents in those classrooms raised a fuss.[24] (See sidebar 6.2.) In this case, teachers didn't realize that they needed to provide an explanation to parents in order to help them understand the rationale behind the new program. One of the reasons that parents objected to the math changes in a New England high school explained earlier was that they didn't understand the reasons for the change. They also worried that colleges wouldn't understand when admissions officials saw their children's transcripts.[25] In both instances teachers failed to realize the importance of explaining their rationales for the changes, and they didn't consider the parents' perspectives when they introduced them.

Even if teachers do not want parents to be included on a school's leadership team or believe that parents should have a voice in making school decisions, every teacher has to deal with parents. What teachers do individually, day to day, has the potential to build support for them and the school. And when parents are given a voice in the way the school is run, they can become the school's biggest advocates and work with the school to bring more parents into the conversation.

PARENTS AS LEADERS IN THE SCHOOL

The mission of the Parent Association (PA) is community building for families of the school.

—Parent Association president

Educational researchers and practitioners understand the key role that principals and teachers play in creating a school with trusting relationships.[26]

Sidebar 6.2

Classroom "Socialism" or a Lesson in Sharing?

Our view: Teachers erred when they collected students' new school supplies as communal stock without warning parents—some of whom then gave kids a poor lesson in sharing.

We're not sure which was worse—the teachers' oversight or the parents' reaction. You be the judge.

What started out as a lesson in sharing for third-graders . . . turned into a crash course on pettiness and lousy communication.

The lesson involved confiscating students' new school supplies and tossing them into a communal pot for use by all students. Why? The idea is "to build a community feeling within the classroom. . . . This also ensures that no child is ever without [supplies,]" said [the] principal . . . explaining that the program is part of the school's "responsive classroom" curriculum.

We don't pretend to be experts in the field of education, but we've never heard of that philosophy. Which isn't to say it's a bad idea. In fact, it sounds wonderfully altruistic, though we have to wonder if the concept isn't lost on . . . 8-YEAR-OLDS!

Apparently so, since some kids came home crying. It's understandable that children would be upset over the loss of school supplies that most kids pick out themselves and that nowadays are imprinted with the images of children's characters and themes.

It's why teachers should have done more than just talk to students ahead of time. The school should have notified parents as well. That way parents could have been brought into the learning process, given the responsibility of stressing the importance of sharing and helping the less fortunate.

Because that wasn't done, parents were taken by surprise. Some were understandably angry.

Unfortunately a few were angry not at being left out of the learning process, but because their children might get stuck with cheaper supplies than the ones mom and dad bought. Had they known about the program, some parents complained, they would have bought generic supplies—as if only *their* child rates a name brand.

That sort of pettiness sends a terrible message to kids: that if you must share, hide the good stuff. Some parents were concerned, like one who e-mailed us, that socialism was being taught to their children. There's a difference, said the parent, between "sharing (voluntary) and confiscation for the common good (socialism)."

OK.

We'll limit our concern to the school's failure to notify parents about an unorthodox program, and parents who then gave children a rotten lesson in sharing.

Source: Editorial, *Bucks County Courier Times*, September 15, 2000. Reprinted with permission.

However, less well understood and studied is the role that parent leaders play in helping to develop these relationships. When parent groups are nurtured by educators, they can play an important role in helping to build trusting relationships among parents and educators. When they are not nurtured, they can play the opposite role.

Every school has a cadre of parents who can be counted on to take leadership roles. They are the school volunteers extraordinaire, the presidents of the Parent Teacher Organization, the members of the school's site management teams, the ones teachers and principals know they can call on in emergencies. The problem many schools have, however, is finding ways to reach out beyond these few dedicated parents to the larger parent community. Schools that harness the power of these parent leaders can build a more engaged parent group.

Schools where parent leaders are not nurtured tend to find it more difficult to reach out to the larger community. In these schools, the parent leadership group is likely to become an exclusive clique, not inviting others to join. However, in a school like Crossroads School, described in chapter 8, parents are not only nurtured by the principal, but they are also involved in leadership-training activities. The principal works closely with the parent leaders building close personal relationships and supporting their efforts. She encourages them to welcome new members to the school community and invite them into the leadership group. As a result, the Crossroads Parent Association takes a proactive role in building an inclusive parent community. Although they have not yet been successful in reaching all parents, especially those of Latino backgrounds, the parent leadership group recognizes the importance of continuing to try.

Parent leaders also can play key roles in helping other parents with less know-how learn how to gain access to school resources and to communicate more effectively with teachers and administrators. They can help model and spread norms of civility. At Crossroads, when it seemed that some parents were abusing their access to teachers' home phones, the parent group—not the principal—assumed responsibility for educating parents about the issue.

PARENTS AND TEACHERS AS STRANGERS: A LACK OF OPPORTUNITIES TO GET TO KNOW EACH OTHER WELL

My relationships with parents are fairly productive. I believe in maintaining contact with parents, so I call parents and keep call-

> *ing them until I make contact. If I don't get them at home, I call at work. If I don't get them in the morning, I call at night. I call until I reach somebody. If I can't reach them by telephone, I go and visit them at home.*
>
> —Joelle Vanderall, teacher, quoted in Michelle Foster,
> Black Teachers on Teaching

Trust requires that we have opportunities to get to know each other in both formal and informal ways. However, in today's hurried world, both parents and educators have limited discretionary time. Parents, most of whom are working outside the home, have to tend to their children's schooling at the same time they are involved in community and civic organizations and must meet the needs of their extended families. Educators, too, in addition to their professional responsibilities, have these commitments—they are parents, spouses, and children of aging parents whom they have to attend to. We all work longer and harder, and our time always seems to be taken up by some new activity. Given this context, finding common times when parents and educators can meet in a relaxed atmosphere is difficult, if not impossible.

In addition to time, changing demographics is another factor that makes parent-educator relationships so hard to build. Young women of European American backgrounds continue to be our primary pool of teachers while our student population is rapidly becoming multicultural and bilingual. Because cultural differences make communication more difficult, they can lead to distrust on the parts of both parents and teachers. Rather than leading to what David Bensman, an educational researcher, calls "cultural interchange" between the school's families and teachers, walls are built, and misunderstandings fester.[27] Rather than seeking ways to learn about the cultures of their students, many teachers believe that such knowledge is irrelevant because teaching is culturally neutral. Rather than bringing the child's culture into the classroom and creating a mirror so that all children can see themselves in the class curriculum, many continue to teach from a European perspective, leaving out or minimizing the contributions of other cultures. Rather than recognizing the importance of helping children maintain their own language while learning English and taking advantage of the opportunities presented when children speak many languages, they resent the difficulties that children who speak other languages bring to their teaching. Rather than asking how different cultures think about children, child rearing, and relationships with teachers, many teachers "read" parents' actions through the lens of their own culturally based beliefs about right and wrong.

Sidebar 6.3

Janice C. Marston, former teacher
Sherwood Heights Elementary School
Auburn, Maine

Reluctant Parents Participate in Arts Integration Days

There are many successful strategies to involve most parents in their childrens' schools. These parents are committed to sharing their child's educational experience and they are comfortable with teachers, administrators, and the school setting. But what about those parents for whom school was an unpleasant experience, leaving them with memories which make it difficult for them to be in their child's school?

Sherwood Heights, a K–6 school of 400 students in Auburn, Maine, instituted Arts Integration Day for several reasons, one of which was to provide a fun, stress-free day when parents not attracted to other school activities could participate without pressure. These days are held three times a year, and, as the title indicates, emphasis is placed on connecting the session content with the curriculum.

Teachers choose what they will do in their sessions, which can be either 45 or 90 minutes long. Projects are divided into either grades K–3 or 4–6, and students are mixed together from throughout the school based on the project they choose. A lending library is available for teachers with suggestions and "how-to" information. Art, drama, music, and dance are represented, sometimes with multicultural, science, or math themes.

One successful project involved the study, design and creation of a basic bridge structure. Another focused on the artist Georgia O'Keeffe: Students created their own works in her style, and learned about her life and career from a guest appearing as a Georgia O'Keeffe look-alike. K–3 students created sidewalk art with chalk, brightening the school area and giving them something of their own creation to share with parents and other students.

Sixth-grade students support these efforts with their Share Center, an ongoing Service Learning Project. Materials come from a regional share center, local businesses, parents, and friends of the school. Teachers may choose from a wide variety of items ranging from art supplies to beads to wooden dowels. Often heard in the Share Center are teacher comments such as "I just had a great idea about what I can do with those," and "I hope I'm not taking more than my share but I have to have all of this for my arts day project."

Letters are sent home asking for volunteers. Since the inception of Arts Integration Days, the number of parents, grandparents, and older siblings participating has grown—as has the enthusiasm of the volunteers. A grandmother asked, "Can I sign up for next time today?" A father said, "I think I'll come to the next PTO meeting. I've never been but now I think it'll be OK." And the older sister sounded a little bit jealous of her sibling: "Nothing as great as this has ever happened to me in school."

Although it is difficult to put specific numbers to a carryover between Arts Integration Day and other school activities, parent attendance has increased at

(continues)

> **Sidebar 6.3** *(continued)*
>
> both the Open House and a Kids Cabaret held each spring. Teachers have found more parents willing to chaperone field trips and to serve as Room Parents. Beyond these considerations, Arts Integration Days add to the student's educational experience, bring parents and others into the school to help and share with both students and teachers, and create a valuable and positive school experience for everyone who takes part.

People need time and commitment to learn about each other's cultures and how to communicate across cultures. And they need both informal and formal opportunities to build trusting relationships with one another. The more different cultures represented in a school, the greater the challenge. Orchestrating the will and time to address this difficult issue can not be just the principal's job. Parents and teachers must help in the important effort of tearing down the cultural walls to build a truly inclusive school community.

Signs of distrust are too easy to see in schools that serve children who live in poverty and represent cultures different from those of the majority of the teachers. Parents who may have had their own bad experiences in schools may question the degree of commitment these teachers have to their children; teachers, unfamiliar with the families of the children whom they are teaching, may make assumptions and judgments about them that are not fair.[28] One story that illustrates this comes from the Chicago schools—a system that serves children from many different cultures. Although it was never fully realized, Chicago School District's plan to "size up parents with checklists" suggests a lack of trust in parents to fulfill their obligations to children. According to an article in *Education Week*, an educational trade weekly, Chicago schoolteachers were asked to evaluate how well parents carried out their responsibilities to their children.[29] For instance, some items were to assess whether "students complete homework assignments and bring their textbooks to school," while others would assess whether they are "eating breakfast, are receiving health care, and are bringing needed medications and eyeglasses to school."[30] Julie Woestehoff, the executive director of Parents United for Responsible Education (PURE), wonders how this assessment will engender a trusting relationship between teachers and parents. She argues that two-way communication about these issues would be a better approach than educators making judgments about parents. The fact that the school district thought issuing "report cards" to parents might be a good idea

Sidebar 6.4

Julie Woestehoff, Executive Director
Parents United for Responsible Education (PURE), Chicago, Illinois

Parent Report Cards in Chicago

In the summer of 2000, the chief executive officer of the Chicago Public Schools announced that Chicago's public elementary schools would begin grading parents on quarterly parent report cards. A national media feeding frenzy ensued when our small parent advocacy organization took exception to this idea, calling it outrageous, insulting, and divisive.

It became clear that the idea of teachers grading parents appealed strongly to a certain group which likes to blame parents for children's school failures. Another immediate and strong response came from parents who wanted to know why they couldn't grade teachers if teachers were going to grade parents. I joined a West Coast radio talk show on the topic, and everyone who called in had a bone to pick with someone—bad teachers, no-show parents, even one deadbeat dad came under attack. Our initial sense that this idea would create more problems than it solved was quickly being borne out.

A look at the actual report card, when it finally appeared several months after the initial announcement, revealed a checklist of standards against which the parent would measured. Most of these standards seemed innocuous enough but a few were quite inappropriate. Teachers would be asked to indicate whether the child came to school prepared, with homework done, and whether the parent participated in school activities. However, teachers were also asked to evaluate whether a parent spent quality time with the child and praised him or her often enough, information teachers are not likely to have and which could be gathered only by infringing on family privacy.

Another set of standards reviewed the parents' compliance with inoculation and other requirements, but included a check-off for enrolling in a voluntary, cost-based state health insurance program. The not-so-subtle message here was that enrollment in this plan was another requirement. Finally, and perhaps most disturbingly, was this standard: "Supports school rules and policies." As a parent who has had to question many school rules and policies during the 16 years my children were public school students, I knew I'd have been marked down on that one every time!

Out in the schools, the report cards found limited acceptance. Most principals chose not to use them, so thousands of glossy trifold "Checklist for Success in Education" now gather dust. Our parent group agrees completely that parent-teacher relationships need improvement. But we also support the National PTA's position that home-school communication must be two way to be effective. Parent-teacher relationships are delicate and easily damaged. Bad parent-teacher interactions have a long shelf life.

We believe that the people in these relationships must have the greatest voice in how they function. As an alternative to the Chicago Public Schools' mandated parent report card, we recommended that each local school council in

(continues)

> **Sidebar 6.4** *(continued)*
>
> Chicago be encouraged to develop its own unique home-school communication plan. We'd like to see school-wide planning so that parents and teachers themselves agree on how they will work together to help our children learn. Students and principals should have a voice, too.
>
> We know that schools would come up with some great ideas. Many already have. We loved the program we heard about at a forum we co-sponsored with a local teacher group—a home-school journal. The home-school journal is a spiral notebook which would travel with the child every day. The teacher writes in it any instructions for the next day, announcements, suggestions, concerns, and even praise for the child and the family. In return, the parent can ask questions, sign necessary forms, or thank the teacher for special help. This kind of approach is far more likely to lead to long-term positive results in student achievement.
>
> Note: PURE is a parent-founded, parent-run group in Chicago that helps parents become more involved in their children's education. PURE offers a wide variety of free informational materials and workshops in English and Spanish. Topics include Local School Council roles, parent involvement, learning standards, and student testing. Organized in 1987, PURE played a key role in writing the 1989 school reform bill that established elected parent-majority councils in each school and continues to bring parents' voices into Chicago education policymaking.

sent a very negative message to all parents in the district—certainly not at all helpful for building trust and respect.

Trusting relationships among all those who have a common commitment to our children will lead to better outcomes for our children. Yet even though this seems like common sense, examples of situations when trust is undermined—both teacher and parent regard each other with suspicion and act in defensive ways with each other—are all too frequent.

As the examples in this chapter have shown, when educators and parents cannot find ways to work together, tensions lead to distrust—and rebuilding trust is much more difficult than establishing it in the first place would have been. Some suggestions for nurturing the fragile beginnings of a trusting relationship appear in the next section.

SUGGESTIONS FOR BEGINNING TO BUILD TRUST AND RESPECT

Building trusting relationships requires constant attention. Every time teachers, administrators, and parents interact, trust can either be built or destroyed

> **Sidebar 6.5**
>
> *Kathy Teel Michel*
> *Formerly Director of Communications, Bettendorf (Iowa) Community Schools*
>
> ## What Do Parents and Educators Want?
>
> Communication plays an integral role for educators and parents. Effective communication requires that parents and educators collaborate, listen to each other, and share information. If we were to diagram this dialogue, it might look something like the following:
>
> Parents say they want to
>
> - Be valued
> - Be heard
> - Have options
> - Know you care about my child and me
> - Be treated uniquely
> - Be listened to
> - Hear positives
>
> Educators say they want
>
> - To be appreciated
> - To be respected
> - Parents to be flexible and compromising
> - To be trusted
> - Consideration
> - To be understood
> - To hear positives
>
> The two sides aren't that far apart, are they? Parents say they want educators who look at the individual needs of each child, encourage positive attitudes in the classroom, and help students work on their self-esteem. Educators want active, involved parents who participate in their child's education.
>
> To that goal, here are perspectives shared by both parents and educators on ways the two can facilitate effective communication.
>
> Parents want
>
> - Affirmation concerning their personal expertise with their child and family
> - Their child to belong
> - To know that educators care about them and their child
> - To be treated uniquely
> - Options
>
> *(continues)*

> **Sidebar 6.5** *(continued)*
>
> - Want to be heard, not just listened to
> - Understanding; they may be dealing with many issues
> - To focus on the "child first," even those with disabilities
>
> In response, educators can
>
> - Provide affirmation by being available for focus groups and forums, by writing newsletter articles, and supporting family centers in schools
> - Support this concept by speaking this philosophy when parents are present
> - Demonstrate this by asking parents and students for input
> - Respect differences by recognizing that everyone does not have the same priority
> - Provide options when they involve parents in the decision-making process
> - Demonstrate that they have heard by providing follow-up calls and letters
> - Acknowledge the issues
> - Promote all children, recognizing differences as gifts.
>
> Both sides need and want to hear positives!

bit by bit. One principal, Cindy O'Shea, says that it is the "little practices" that count most. Every time a parent and teacher meet before school or in the local market, every time a principal speaks with a parent, is an opportunity for trust building—for letting the parent know that the teacher or the principal is committed to their child's best interests.

On the other hand, every time that a teacher puts off a parent or a principal is brusque, trust is diminished. During the rush of the daily school activities, a parent's request for a teacher's attention might be difficult to accommodate, but the way in which the teacher suggests another time will communicate to the parent either "I really want to talk to you, but this is not a good time" or "Go away. Can't you see I'm too busy to talk to you now?"

The building blocks of trust are few: a belief in the competence of each other, a belief in the reliability of each other, open relationships and communication, and a recognition of mutual concerns about each child. Because Greenbrook Elementary School and Crossroads School are currently doing much of what is suggested below, they have built positive relationships. Full profiles of these two schools follow in chapters 7 and 8. Chapter 9 includes many other practices to add to the suggestions that follow:

Share Perspectives: Two-Way Communication and Active Listening

> *[L]istening to parents . . . needs to be seen as a crucial element in any attempt to improve home/school relations.*
>
> —Janet Atkin and John Bastiani, *Listening to Parents*

An absolute first step for building trust and respect is understanding and gaining appreciation for each other's perspectives. Somehow we must find ways to try to understand how we each experience the world. What is it like to be a teacher of over 100 students every day? What is it like to raise three children alone on a minimum salary? Why do you think teaching reading solely through phonics is a good idea? Why do you think block scheduling is a good idea? What have you learned about my child? What do you know about your child that can help me teach her?

Schools can do a lot to encourage this type of communication. They can invite parents in to tell teachers what they know about their children. Some schools do this at the beginning of every school year. Some teachers, like Lori Woods at Greenbrook School, make home visits. They can ask parents for their opinions about new programs. They can have social gatherings where parents and teachers share stories about what and how they learned when they went to school.

Re-Define Professionalism

> *. . . professionalism [must be] redefined. To be a professional educator would include the ability to relate to and team with a variety of people and organizations—such as parents, citizens, and social agencies."*
>
> —Mary Henry, *Parent-School Collaboration*

If we think of the relationship between educators and parents as one of overlapping spheres of influence, as suggested by Joyce Epstein, an expert on school-family relationships, then tension is bound to occur in the areas of overlap.[31] Then teachers must rethink their professional role to include working *with* and learning *from* parents, recognize that communicating with parents is an essential responsibility, and take the lead in reaching out to parents to negotiate the troubled waters of those areas of overlap. If they do so,

Sidebar 6.6

Nancy Goldberg, Teacher
Consolidated School, Kennebunkport, Maine

Communication Is Key

Parents have been an integral part of my kindergarten program for 12 years. Open communication with the parents of my students is a key component of a successful kindergarten program. Starting a full nine months before classes start in September, I begin a dialogue with parents that will extend throughout the kindergarten year. These communications include

- Kindergarten registration: Parents register their children in December of the year prior to kindergarten. At kindergarten registration the special education staff, school nurse, and I met with parents of the incoming students to review a development questionnaire, gather health information, and listen to concerns parents may have.
- Kindergarten screening: The following May, we administer a kindergarten screening. At the same time, parents meet the primary school principal and guidance counselor to hear an overview of the kindergarten screening, the district's kindergarten and early kindergarten programs, and what to expect the first few weeks of school.
- Newsletters: In August, I send families a welcome newsletter with details of the first few weeks of school. During the school year, I write a monthly newsletter for parents to inform them of upcoming classroom themes, projects, field trips, and special events.
- Sneak-a-peek: The week before school officially begins, the kindergarten children meet their classmates and me in the classroom, while veteran parent volunteers welcome new parents at a reception in our school cafeteria. The parents discuss the school's parent group, ways to be involved in their children's education, and volunteer opportunities. Veterans give a tour of the school to new parents.
- Information night: This meeting, scheduled during the second week of classes, allows me to inform parents about our curriculum, daily schedule, school staff, and parent-volunteer program.
- Volunteer training: The objectives of this late-September meeting are (1) to put parents at ease with their role as volunteers so that they can relax and enjoy it, and (2) to foster understanding of my objectives for the classroom and my teaching and management styles.
- Project letters: At the end of each month, I write a project description letter that goes home to parents with a big bag full of the children's monthly projects. The project letter also gives parents suggestions for activities to extend learning at home.
- Home folders: On a weekly basis, each child carries home notices to parents and homework assignments for completion the following week.

(continues)

> **Sidebar 6.6** *(continued)*
>
> - Conferences: In November, I meet with parents of students to discuss the results of the preschool screening, the initial reading and math assignments, and classroom progress to date. I explain to parents specific activities they can do with their children at home to enhance learning. Parents and I confer for a second time in April.
> - Report cards: I complete report cards assessments in January and June. The assessment informs parents about various aspects of student progress through the school year.
> - Telephone conferences, ad hoc conferences, and written notes: Supplementing the parent-teacher conferences in November and April are unscheduled notes to and from school, telephone calls, and additional conferences conducted on a needed basis.

parents will more readily recognize that they must work together with teachers in the best interests of their children. They will begin with the belief (until proven otherwise) that teachers have the best interests of their children at heart. They will approach teachers with respect believing that differences can be bridged.

All educators—leaders in schools and in professional organizations as well as those of us who prepare new teachers—have a role to play to see that this happens.

Learn About Each Other's Cultures

> *We began by defining "cultural interchange" as the process by which members of groups with different traditions, values, beliefs, and experiences gained a greater degree of mutual understanding.*
>
> —David Bensman, *Building School-Family Partnerships in a South Bronx Classroom*

As communities become more diverse, all of us—educators, parents, and community members—must learn about each other's cultures. We each see and make sense of our world through our cultural lenses. If we understand the world only through our own lenses, we are sure to misunderstand each other's intentions and motives. Researcher David Bensman suggests that we strive toward a "cultural interchange" in our schools, where teachers, parents, and community members engage in activities to learn about each other's worlds.[32]

While teachers must understand the cultures of their students, parents, need to learn about the cultures of their children's classmates. The conversation at a Greenbrook Kindergarten Parent Journal Writing meeting turned to this very topic. The group consisted of mostly white parents and one African American parent. One parent explained that, because she had grown up in a community where everyone looked and talked as she did, she had very little experience with people from different backgrounds. Now her child was in a class with African American children as well as with children from India, China, and Mexico. She said, "I still feel awkward with people who are different from me. I am so pleased that my child is learning from a very young age about people who have different backgrounds. When she grows up she will be able to relate in ways that I can't." Over time this informal discussion group developed a sense of trust in each other and gave these parents an opportunity to confront this issue. More opportunities like this are needed.

Create Formal and Informal Opportunities for Getting to Know Each Other

> *It's hard to create relationships with the parents of 2000 kids. Back-to-school nights don't do it. The best results come from advisory systems or very small "houses," where parental contact is facilitated by the fact that a student may work with the same teachers for several years.*
>
> —Rick Lear, Coalition of Essential Schools,
> quoted in "Essential Collaborators:
> Parents, School, and Community"

Educators and parents in trusting school communities have many different formal and informal opportunities to get to know each other well. These interactions happen almost naturally in small schools and communities but are difficult in larger schools. Schools with thousands of children make this impossible. Teachers who teach over 100 students each week will find it impossible to know all of them and their families well. Schools, then, need to find ways to create smaller communities of teachers, students, and parents so that knowing each other isn't an impossibility. Some schools have done this by creating houses or schools within schools.

> **Sidebar 6.7**
>
> *Jean Konzal, The College of New Jersey*
> *Ewing, New Jersey*
>
> ## Claire's School[*]
>
> Tucked into the hills of southeastern Vermont is the Westminster West Elementary School. Originally a one-room schoolhouse for children ages 5 to 8, it now houses two classrooms—one serves children 5 to 7 years old and the other children 8 to 10. Claire Oglesby, the early primary teacher, a transplant from New York City many years ago, has been teaching children here since 1972, first in the one-room schoolhouse and now with a colleague. This school is her creation. It reflects her strong beliefs about how children learn best—in a community of caring, supportive, and challenging adults and peers. Co-author Jean Konzal's son, Gregory, attended this school as a second and third grader in the late 1970s. As parents, Jean and her husband, Bill, were quickly drawn into the small school community. There was an open invitation to participate fully in the life of the school. The school was always abuzz with parents. When one walked into this first- through third-grade classroom, it was common to see children working together in groups, parents working with children, parents talking with each other or with Claire or her aide.
>
> Claire knew all of the families well and families trusted her with their children. Children were involved in long-term challenging projects, and families were drawn into the activities. Children and families from different backgrounds (some Vermonters for generations, others recently arrived "flatlanders" from New York City) worked side by side, forging new friendships and developing respect for each other.
>
> Parents met in the evenings in the school for a variety of reasons—to work on projects, to talk about parenting, to share experiences. Parents were recruited to contribute to the school in a variety of ways, depending on their skills, interests, and resources—to lead field trips to where they worked in the community (farms, sawmills, apple orchards, and cider mills), to help with school projects (building, cooking, constructing), and, as in the case of Jean, to seek funds for additional programs by writing grants.
>
> The school provided a community gathering place. It was from this experience that Jean developed her image of what a school might be.

---

[*] For further information, see the video *The World in Claire's Classroom* by Lisa Merton and Alan Dater. (Distributed by New Day Films, 22-D Hollywood Avenue, Hohokus, New Jersey 07423.)

Crossroads School in Manhattan was deliberately designed to be small—and to share a building with other schools. Each student is part of an advisory group of about 12 students and 1 teacher. This teacher is responsible for guiding the students in her or his group. This means keeping tabs on how the child is doing in all classes, identifying problems early and heading them off, and communicating with parents. All parents have their child's advisor's home phone number.

In addition, the parents' association and the school administrators plan and sponsor many different opportunities for parents to get together with parents and for parents and teachers to get together. One such event is the picnic at the end of the school year. Students who will be new to the school in the fall are invited along with their parents to meet the current students, parents, and teachers. Many parents remember this picnic as a time when they were made to feel a part of the school community.

Speak to Others as You Would Want Them to Speak to You

> *This parent was very threatening, yelling at me that people get fired for doing less than this. He was in my face and I kept saying this isn't appropriate. I was scared.*
>
> —Robin Campbell, teacher, quoted in Colleen Pohlig, "Schools Spelling Out Need for Parental Civility"

When under stress, anyone, unless well trained and practiced in good communication skills, tends to let anger or frustration speak. And once an angry word is spoken, the damage has been done. Many teachers report a sense of "being burned" by the harsh words of angry parents—to the point of not wanting to put their guard down again for fear of being attacked again. This was a recurring theme in the Springfield story told in chapter 4. On the other hand, teachers who dismiss parents' requests or speak to parents in patronizing tones will find it hard to rebuild broken trust, as the South Wellington story in chapter 3 shows.

Teachers and administrators need to develop their conflict resolution skills so that when they are confronted by an angry parent acting inappropriately, they can defuse the situation. Preservice teachers should learn how to resolve conflict in positive ways in teacher-education programs. Once they begin teaching, they should have ongoing staff development to sharpen and develop these skills further. Parents also can benefit from learning better ways to manage conflicts because these skills will be very useful when dealing with their own children, not just their children's teachers. Parent organizations can play a key role by providing workshops about how to advocate for children's needs in ways that are effective and civil.

Open communication based on trusting relationships—where motives are not questioned, where honest disagreements are explored, where people listen and learn from each other, where it's safe to say what one really thinks—can lead to better educational decisions for all children.

Promote Shared Leadership

> *Most people view principals as leaders. But others in schools— teacher, librarians, guidance counselors, parents, and students— also want to make things that they believe in happen. The kind of school I'd like my children to attend is one in which everyone gets a chance to be a leader.*
>
> —Roland Barth, *A Personal Vision of a Good School*

School leadership should not be in the hands of just one person—the principal—but should be shared with teachers, argues scholar Roland Barth, who calls to mind the image of the point bird of a flock of geese falling back and giving others a chance to lead.[33] There are good reasons for extending the idea of leadership to parents, as well.

At different times and under different circumstances, it may make sense for a teacher or a parent to take the lead rather than the principal. In order for this to work, teachers and parents need to be involved in planning school change efforts and need to have an opportunity to develop skills necessary to lead. Some schools, recognizing that parents on advisory boards need to develop leadership skills to enable them to participate fully and to take leadership positions, offer them training. Principals must be given the support they need—professional development and mentoring—to help them develop the skills necessary to create trusting relationships in their schools. At the least, there should be an equal value placed on relationship building and increasing test scores. In fact, studies suggest that schools with trusting relationships are more successful.[34]

Sidebar 6.8

Larry DeBlois
High school English teacher (retired)
Augusta, Maine

The Essential Advisor

If you have a problem at your child's school, whom do you call? At my school, if you want something done, you call the advisor. Each child has an advisor. There is one advisor for groups of 12 students at the same grade level. The advisor is the one who knows what's happening with your child, who can find the answers you need, and who is accessible. You want answers, call the advisor.

The advisor is an adult guide for the journey to graduation and is selected for the child when s/he enters the school, in the same way a locker and classes are assigned. Since our school received students from four different elementary schools, each group was made of students from each of the towns. No attempt was made to place friends in the same group. The plan was that all these disparate members would learn to be friendly, to get along, and to work together on projects that benefited the school and community. Along the way, they learned to be friends with all the members, even those not on their social level.

Each teacher at Maranacook Community School is an advisor; advising is included in the contract as one of the prime duties. (The other prime duty was teaching.) As advisor, I met with my advisees daily in homeroom, we planned activities, days out, and community service projects. I listened to their problems, helped them sign up for courses, interceded with them with teachers, and advocated for them in disciplinary meetings (very few actually). I was the adult at school most concerned with their progress. I was the adult at school who was most connected with them and had the most influence with them.

If a teacher experienced some difficulty with one of my advisees, that teacher wrote me a note describing the incident. I took the note to the advisee and asked, "What's going on?" We talked, I suggested more appropriate behavior, and then I reported to the complaining teacher. If the incident was a repetition of several other infractions, then it was my responsibility to call a disciplinary meeting. If, in addition to frequency, the incident was one of the serious four (drugs, alcohol, disrespect of persons or of property), then I called the parents and set up a meeting with them in attendance. I acted as the advocate for my advisees—defending, explaining, and, at times, agreeing on punishment. (If my advisees got in a fight, I got them to explain their version of what happened. But all fighting in school had to result in suspension and that was the punishment. I concurred. If I hadn't, we would have devised another, more appropriate punishment. I always represented what was best for my advisee.)

But talking to my advisee was generally enough to see change in behavior. They knew I was looking out for their best interest. No one disappeared in our school. Everyone had an advisor and advisors checked on things. The students knew that. We weren't spies, though; we were advocates.

(continues)

> **Sidebar 6.8** *(continued)*
>
> My group always liked to go out for dinner as an activity. As long as the group agreed to the activity, and we found parents to transport us, I went along with most ideas. (Paintball wasn't one of them.) We also went to the beach, and climbed to the Tuckerman's Ravine hut, and camped out at Acadia National Park to mention a few activities that we planned. We also brought in wood for an elderly community member and made Christmas cards for each of the occupants of a senior-housing unit.
>
> I'm still in touch with many of my advisees, years after they've graduated. Working with them outside of a classroom was different and did take some extra work in planning. But I had the privilege of watching them mature through the years and being a part of their lives for six years. I gave them their diplomas at graduation and then watched them move into the adult world. I can't imagine a school without a similar system.

Value Small Practices

> *We're only as good as that last interaction. Everything can get wiped out by one mistake. We have to continuously be proactive.*
>
> —Pat Holliday, Principal

And finally, everyone in a school—principal, teacher, secretary, custodian—must recognize that every interaction they have with a parent provides an opportunity either to build and sustain trust or to lose it. There must be an overt and articulated recognition that an important part of every person's job is to make each and every contact with parents and community members count.

The next two chapters profile two schools that have made major strides in building trust with their families and communities. Their efforts show that building trust is an ongoing process. Greenbrook and Crossroads educators continuously think about and work with all of their families. These schools are identified—without pseudonyms—and are presented certainly in a positive light, but also with a discussion of the tensions that are yet to be resolved. Teachers, principals, and parents talk about what they do with each other and want to share their successes and their struggles with others. They have stepped out from behind the typical public relations approach that hides problems and instead choose to provide an honest reflection of who they are for public scrutiny.

The new paradigm of parent–school relationships is characterized as a community of learners. Learning can take place only when all participants celebrate their successes and openly and honestly confront their problems. Their willingness to present themselves so honestly here is a sign that these schools are moving toward the realization of a new paradigm of home–school–community relationships.

Part III

MOVING CLOSER TO
THE NEW PARADIGM:
PROFILES AND PRACTICES

7

A PLACE WHERE EVERYBODY KNOWS YOUR NAME: GREENBROOK ELEMENTARY SCHOOL

> *Greenbrook is a school where "everybody knows your name," starting right at the top with the principal. Not only does she know every child's name but the parents as well. Our school is built on a strong sense of community. And that community nurtures me on a constant basis through the students, the teachers, the principal and the other parents.*
>
> —Shari Rothstein, parent, Greenbrook Elementary School

It is 9:30 on a Wednesday morning in May in the Greenbrook Elementary School (GES) library. Nine parents of this year's kindergarten class, dressed in white tops and black slacks or skirts and holding black three-hole binders, are perched on stools in front of an audience of parents of next year's kindergarten class. They begin to read from their scripts:

Mary: As I think back to the spring day in May when you had your kindergarten orientation, I can remember how excited you were. I was a little nervous.

Betsy: I realize that our "special" time together, our preschool years, are coming to an end. My heart was breaking this morning as you grabbed me around the neck, held me tight, and begged me to take you home. Part of me wanted to scoop you up and head for the door.

Tracey: I'm so glad I had my dark sunglasses on so you couldn't see that I was crying. I don't know how I'll get through the year.

Miriam: You have been so excited for today.
Eleanore: I was sent to the library. I knew I just had one hurdle to cross—the kindergarten journal project performance!
All: We all cried.[1]

The script the parents read from is constructed from journal excerpts written during their children's kindergarten year. The parents go on to share their hopes, fears, joys, and surprises as they watched their children grow and learn during this, their first year of school. Now in its fifth year, this parent-to-parent sharing of experiences has become a tradition at Greenbrook. The project all began in the summer of 1997 when the GES Site Council met to complete the school's strategic plan for the coming year. In an effort to find a way to help make parents new to Greenbrook School feel part of the community, the council came up with the idea of inviting parents to keep journals about their children's kindergarten experiences.[2] This project, however, is only one way the school reaches out to parents.

THE SCHOOL

Greenbrook Elementary School, a K–4 school of over 300 children in a rapidly growing New Jersey school district, is in a predominantly white middle-class community of modest older homes and newer, more substantial homes. Data from 2000 to 2001 show that while a large majority of the school population is European American and English-speaking (75 percent), there is a growing presence of other ethnic groups: Asian, mostly Indian and Chinese (16.5 percent), Latino (3.7 percent) and African American (4.8 percent). Other than the principal and the school librarian, who are African-American, and one teacher, who is of Asian descent, the faculty is composed of persons of European American descent. During the 2001–2002 school year this number increased slightly: Four members of the professional staff are African American. The school district has the state's highest rating in terms of economic security; however, about 5 percent of the school's families qualify for free or reduced lunch, and, according to the principal's estimates, an additional 5 percent face other challenges that make life more difficult for them. These numbers may change in the future since, due to the rapidly growing population, the school district is building a new elementary school and making changes as to which school's neighborhood children will attend.

> **Sidebar 7.1**
>
> ### From the Principal's Desk: *Greenbrook Community News*
>
> October/November 2001
> Dear Parents and Guardians,
>
> The recent tragedy at the World Trade Center has touched us all deeply. I am very thankful that our immediate school community is intact, but I am aware that many of us have been personally affected by loss nonetheless. As we try to begin to return to a sense of normalcy, if the school resources can be of help, please call on me. Mrs. Myers, our school counselor, and I are available to assist you.
>
> Be assured that our mission here at Greenbrook School continues to be focused on providing all of our children with a safe, nurturing, and academically challenging experience each day. Your children are our future and the hope for a peaceful world.
>
> The faculty and I enjoyed meeting you Monday evening at Open House. I hope you enjoyed having a chance to meet the staff, hearing about your child's day in school and learning about our programs. I also hope that many of you will be able to join us for the rescheduled Welcoming Picnic, on Saturday the 6th of October from 1–4 P.M.
>
> This casual gathering will give parents an opportunity to meet other families and strengthen our school community.
>
> Sincerely,
> Patricia Holliday

The school, under the leadership of principal Pat Holliday and with the support of the teachers, is a member of the Coalition of Essential Schools, a national network of schools bound together by a common set of principles that guides the work of the schools and plays a leadership role in New Jersey's Coalition network. Greenbrook is also a member of The Professional Development School Network of The College of New Jersey, where it has played a major role in developing the college's new cluster student teacher model—partnering with a college supervisor to mentor a group of students each fall semester.

GES has the reputation of being a place where children and families feel welcomed and "at home." As Carol Desmond, a parent, has written: "Greenbrook School is a 'parent friendly' community . . . that welcomes, demands, and thrives on parent involvement . . . a community that parents are proud to be an integral part of."[3] The school offers a wide array of both formal and informal opportunities for parents. These activities allow parents and teachers and principal to get to know each other well. The principal knows each and every child and family and is easily accessible to them. Mary Henry, a noted educational researcher, has observed that it is the informal opportunities to develop relationships that help to create a school culture open to parents.[4]

Visitors to this one-story, circa–1961 building, with a central core and two classroom wings on either side, see hints that this school might be different from other schools. First, the walkway leading to the school is landscaped with butterfly and perennial gardens carefully tended by parents. Each spring during the Math, Science, and Technology Day/Night, children, parents, and teachers plant, weed, and tend these gardens. During the morning drop-off time, the school custodian warmly greets parents and children at the car drop-off point while the assistant principal and a paraprofessional welcome children as they tumble out of the buses. Visitors sense an informal, open, and caring atmosphere as they follow the children into the school where they spread out to their classrooms. Kindergarten through grade 2 children circle the open and sunken multipurpose room to the right, and third- and fourth-grade children circle to the left. The sunken, multipurpose room adds to this sense of informality. This space, open to the corridors that circle it, has a stage at one end and two four-step-down entrances on the opposite end; it is used for gym classes, assemblies, and lunch. Before-school and after-school programs are also found here. During the day visitors are likely to see gym classes in motion or a rehearsal for a play on the stage.

Parents are gently but firmly reminded to sign in at the building office, but without the offputting notices found in some schools.[5] The school's interior, somewhat worn around the edges, showcases children's work. Children's projects are hung with care, but not in a stiff and formal display as is found in some schools. The message here is that this is a place where children are working, learning, and creating. On any given day visitors will see parents working in classrooms as volunteers, reading with children seated on the floor in the school's corridors, or working on school or class projects in the faculty room.

The school faculty room is the hub of the school—a place where teachers, staff, parents, and students gather in small groups or individually to chat, to work, or to plan together. The faculty room is furnished with tables set up in a large square and also includes a work area with copiers and curriculum materials, a kitchen area (used often by parent volunteers cooking or baking with children), and the only faculty rest room. The room hums with activity, sometimes hosting parents working on a project for their PTO and at other times hosting the ESL teacher and a small group of children or a student-teacher seminar. (Space at this school is at a premium.) A big whiteboard acts as the central communication tool for the school. Daily messages are posted here: "Who has an extra . . . ?" "I have extra. . . . , Does anyone need any?" " . . . 's mother/brother/sister is in the hospital." "Don't forget to . . ." In many ways this room is the metaphor for the school, a place open to all, a place where everyone is welcome, where people laugh, let off steam, and work hard.

Not only is the school a busy place during the day, but it hosts before- and after-school programs for working parents, and there always seems to be some kind of evening event. The before- and after-school programs, serving more than 50 students, is sponsored by the district's Community Education Program and directed by the school's basic skills teacher, Moses White. Each month the program addresses a different educational theme. For instance, September was "Bugs and Insects" month; October focused on the performing arts. Early in October 2001 the program hosted a "Lights On" celebration. Parents were invited to join their children and the after-school staff in a late-afternoon and early-evening celebration of the performing arts. Children danced, sang, and played instruments for their parents. The program is in the process of seeking accreditation from the National School Age Child Care Alliance.

Central to all the activity in the school is the principal, whom teachers and parents call Pat. Parent Carol Desmond notes that "Greenbrook School is . . . a community . . . where the principal sets the tone." Pat sets this tone by greeting parents by name and talking knowingly with them about their children. Miriam Lally is only one parent who appreciates Pat's efforts:

> One of my earliest memories of Pat Holliday was that she greeted me by name one day as I entered the school. I was a new parent at Greenbrook. It was early in the school year, and I don't think she had met me more than once. This really impressed me since there are so many students and parents in the school. Small things like this help create a small community feeling in the school and assure the parent that the principal is closely connected to students and family.

Another parent, Shari Rothstein, recalls:

> Whenever I have needed to speak to our principal, she has been readily available. My phone calls get returned promptly and my requests for meetings are

granted quickly. I never feel rushed when speaking with her. She listens to my concerns and takes notes. To me, that says she cares. I am partners with the principal, partners to make the school the best place it can be. I can call on her and she can—and has—called on me to help. When she wanted parents to come to a Board of Education meeting, she called us. She cared and respected us enough to call us.

Pat and her teachers demonstrate their commitment to parents in a number of other ways. For instance, Pat attends numerous nightly parent meetings. Following her example, teachers also attend many school-related events. One teacher said, "When we attend evening meetings, parents see this as an act of good faith. We're willing to put in extra time for their children. They see that caring for their children is a mutual thing." Pat recognizes that involving parents in their children's schooling is of primary importance and is much more than a series of programs. According to Pat, "Parent involvement is about creating a sense of community, of letting each and every parent know that my staff and I care deeply about their children. This happens each time a teacher and a parent talk with each other, each time I greet a parent in school, each time a parent has a concern and feels free to bring it to my attention. It is not a series of canned programs or a written policy that lies dormant in some document."

Pat's philosophy provides the context for Greenbrook's parent-participation programs. They are not "parent-involvement projects" attached to a school, as educational scholar Ernest Boyer would say, "like barnacles on a ship," but instead are nested in a culture that encourages and nurtures parent engagement.[6]

PARENT PARTICIPATION IN SCHOOLWIDE ACTIVITIES

By creating a number of schoolwide activities, Greenbrook Elementary School invites parents into the learning community as participants and spectators, as volunteers, as learners, and as decisionmakers.

Parents as Participants and Spectators

The Welcome Back Family Picnic gives families a chance to meet their children's teachers as well as families of other children. Parent Carol Desmond

> **Sidebar 7.2**
>
> ## A Typical Year at Greenbrook
>
> *2000–2001 Calendar*
> A review of the Greenbrook calendar demonstrates the variety of ways that parents are invited to be part of the school. Principal Pat, the teachers, and the school's parent leadership know that "one size does not fit all," and they try to provide a wide variety of opportunities to accommodate as many parents as possible.
>
> First Tuesday of Month/8:00 P.M.: Parent-Teacher Organization
> Third Wednesday of Month/7:30 P.M.: School Site Council
> First Thursday of Month/ 7:30 P.M.: Kindergarten Parent Journal Writing project meeting
> Friday, September 8, 2000: *Welcome Back Family Picnic*
> October 5, 2000: *Family Read-Aloud Night*
> October 11, 18, 2000: Operation Bookworm Volunteer Training
> Thursday, November 16, 2000: Mrs. Sisko's class' storytelling festival
> Thursday, December 7, 2000/7:00 P.M.: PTO-sponsored Holiday Craft Event
> Wednesday, February 7, 2001: World premiere of Mrs. Dahlberg's class video animation project
> Sunday, January 21, 2001/6:30–8 P.M.: PTO-sponsored family ice skating event
> Saturday, February 10, 2001: PTO-sponsored Parents' Valentine's Day Dance
> Thursday, February 15, 2001: PTO-sponsored kindergarten–grade 2 talent show.
> Wednesday, February 21, 2001: The Write Stuff Workshop for Parents
> Thursday, March 15, 2001: PTO-sponsored third- and fourth-grade talent show
> Sunday, March 25, 2001: PTO-sponsored family bowling event
> Wednesday, March 28, 2001: Math, Science and Technology Night
> Wednesday, April 25, 2001: K–1 Child Development Workshop
> Wednesday, May 9, 2001: PTO-sponsored Staff Appreciation Luncheon
> Tuesday, May 15, 2001: Author's Night
> Thursday, May 24, 2001: PTO Walkathon

explains: "This is an informal get-together on a Friday evening, held outside on the school grounds. Each family brings their own dinner. For a new Greenbrook family it is a nice informal event where you can meet and mingle with other Greenbrook families, and the principal, who sits at the entrance and formally welcomes everyone individually as they arrive. It's a small event, which takes very little organizing, but it makes new parents and children feel welcomed into the Greenbrook school community."

Author's Night, sponsored by the School Site Council, gives children the opportunity to showcase their writing and illustrations. Children read books they have written in front of small audiences of their peers, teachers, and parents. In their kindergarten journals many parents commented on the pleasure they felt when their children participated in this event.

As in most elementary schools, the music teachers present winter and spring evening concerts for parent audiences. Each class has an opportunity to shine. And, of course, parents turn out in droves to see their children perform.

Every Friday morning the whole school gathers for a brief school community assembly. This is a time for birthdays to be acknowledged and for children from each of the classes to show everyone what they have been learning. Classes write and perform short skits or plays, recite poems, or sing songs about what they have been studying in school. The assembly program was created as a way for each class to showcase its students and extend the classroom learning experience. In the process of preparing for an assembly, students also develop poise as they practice their oral presentation skills. For example, one class of second graders shared their study of animals and their habitats with the rest of the school; they used the information gathered from researching several different types of animals to create murallike illustrations as a background. These short presentations are child-centered performances, not Cecil B. DeMille productions. Children sit on the floor of the multipurpose room while parents line the corridors surrounding this sunken space to watch their children perform.

Parents as Volunteers

The Parent-Teacher Organization (PTO) is very active. It produces *Greenbrook Community News*, a bimonthly newsletter for the whole school community; raises money for school needs through events, such as a "Walkathon"; and organizes family events, such as "Family Bowling," and school events, such as arts appreciation programs. In 2000–2001, for example, the PTO sponsored the Pilobolus Dance Company residency, which included both workshops and performances. The PTO is also the organizer of K–1 and 2–4 talent shows and recognizes the efforts of the Greenbrook faculty and staff through events, such as the Staff Appreciation Luncheon.

Operation Bookworm was created by the reading specialist in the school, Ellen Gordon, and her colleague, Eileen Zweig, another reading specialist in the district. The volunteer program brings parents, community members, and high-school students into Greenbrook Elementary School—and other elementary schools in the district—to provide K–2 students with extra reading experiences.

Volunteers are trained to work one-on-one with selected kindergarten, first-, and second-grade students, helping them to learn how to read. Together they read and talk about books, do follow-up activities, make up and write stories, and play alphabet games. Volunteers learn special reading techniques by attending two workshops. These techniques are based on researched-based early childhood reading strategies. The workshops give volunteers the practice, strategies, and confidence needed to read with children. Parent Carol Desmond says, "Parents who have participated in this program find it extremely rewarding, and the 'bookworm' students seem to thrive on the individual attention. One of the students that I work with begs to be read to whenever he sees me around the school, so I know he loves the time we spend together, and I can see the progress in his letter recognition and beginning reading skills."

Family Read-Aloud Night is sponsored by the School Site Council and a faculty committee. Principal Pat describes the purpose in a note to the staff in the weekly staff newsletter, *The Wave:* This event is a "low-key evening celebrating families reading together. Mr. Zadek [a parent] did a wonderful job sharing a few original poems and telling the story of a haunted house, and then we moved easily to the classrooms for more intimate stories. As you know, we have ongoing reading goals in our school plan, and an evening like this helps us strengthen this area. I especially enjoyed watching the parents interact with their children as they read."

Parents as Learners

The Kindergarten Parent Journal Writing project, introduced at the beginning of this chapter, typifies the kind of opportunities for parent involvement that are available to Greenbrook parents. Parents of kindergarten children are invited to keep personal journals of their children's school year and to attend monthly meetings to share excerpts from their journals and to discuss issues of common interest that emerge from their journals. They use journal excerpts to prepare a readers' theater script and then perform the script at the orientation for the parents of next year's kindergarten children.

The journals give parents a way to explore issues of child development, parenting, and the school's academic program. At the monthly meetings the parents (mostly moms) can discuss their concerns and hopes with other parents. In the past four years of the project, for instance, parents have talked about their fears that their children would not make friends and would not learn to read. They have shared their own feelings of personal discomfort when they interact with people who are different since most grew up in predominantly homogeneous white communities. Yet at the same time they have said how pleased they are that their children benefit from living in a much more diverse community. Parents also have talked about how their children are being taught to read and write. For instance, although many parents initially expressed concerns about "inventive spelling," they began to think differently when they saw how confident their children were becoming as writers. They could see the value of inventive spelling as children's writing began to become more recognizable, that is, when their children began using correct beginning and ending consonants.

The journal meetings also became a place for parents to talk with Pat about concerns they had about school. In fact, one of these conversations during the first year led to the idea of having parents use their journal excerpts to create a presentation for the incoming kindergarten parents. The parents said that the orientation they attended hadn't really addressed their concerns, and they engaged the principal in a conversation about the kindergarten orientation. While the orientation provided useful information about the school's services and programs, it hadn't really cut to the heart of their concerns—concerns reiterated by the following excerpt from the 1998 script:

> Would there be someone there to take care of all the little things that I had been there for up until now? Would his teacher get to know the real Wally? Would he make friends? Would he make friends with nice kids? How would he adapt to peer pressure? Could he sit still for a whole school day? Could he learn to raise his hand and wait for his turn before he spouted out all of the interesting thoughts, stories, and ideas that he has to share? Could he open his thermos, keep his backpack in order, tie his shoes, button buttons, take charge of his things, etc. etc. etc.?[7]

Pat listened and then invited the parents to provide the missing piece of the orientation—parent to parent. Thus began the four-year-old tradition of parents publicly performing their readers' theater scripts. Now every kindergarten parent orientation begins with a performance of the script written by the previous year's kindergarten parents. Together, parents experience the emotions they share as they prepare for this first of many milestones in their children's journey through school. Many cry as they prepare to see their children enter kindergarten—a sure sign that children are growing up. In fact, many parent performers think that the success of their performance can best be measured by the number of parents in the audience who reach for a tissue.

The journal project and the performance exemplify many of the qualities of a new-paradigm parent involvement program: parents as active learners and educators as active listeners. Through authentic learning experiences, parents act as researchers of their children's kindergarten year, constructing knowledge with others about children's social, emotional, and cognitive development. As parents watch their children learn to read, write, and think mathematically, they also learn about the school's approach to teaching and learning. The principal invites parents to speak honestly and openly about school policies and practices. She wants them to tell her how these policies affect their children and families—and she listens carefully, explains, and invites suggestions for changes. Parents feel not only that Pat listens but also that she really *hears* them and will act on their concerns—and she does.

This type of program supports parents as lifelong learners. One of the four groups of parents that have participated in this program, for example, continued to meet in each other's homes after their children moved on to first grade. Because they continue to write in their journals as their children progress through the grades and to meet with parents, they continue to learn—and to provide support and encouragement to each other as they face new challenges as parents.

The Write Stuff Workshop for Parents, sponsored by the School Site Council and presented by teachers Ellen Gordon (reading specialist), Janet Gill (third grade), Alisa Algava (fourth grade), Beverly Dezan (second grade), and Nancy Sears (second grade) offers parents an opportunity to learn about how children learn to write. Greenbrook teachers discuss the writing process and how parents can help their children at home. One parent who attended the workshop commented, "I learned important skills that helped me to help my child with their writing." Another said, "This workshop helped me take the 'pain' out of writing for my son!"

The K-1 Child Development Workshop, also sponsored by the School Site Council and presented by teachers Lois Margolis (kindergarten), Lori Woods (multi-age classroom), Ellen Gordon (reading specialist), and Erica Cho (first grade), helps parents think about how young children learn. In the in-house faculty and staff newsletter, *The Wave*, the assistant principal, Jan Bozowski writes, "I'm always tickled to watch teachers realize what a great amount of knowledge they have about how children learn. I think the parents gained valuable insights that night." Both this workshop and the Write Stuff workshop showcase teachers' knowledge and expertise and provide parents with new knowledge that fosters their confidence in their children's teachers.

They can see that teachers have thought a great deal about what they do in their classrooms and are worthy of their trust and respect.

The Math, Science, and Technology Day/Night, another effort of the School Site Council and coordinated by one of the teachers, Erica Cho, offers families a variety of math, science, and technology workshops. For example, parents and children could choose from such offerings as the "Take-Apart Room," "Rocket Launching," "M & M Math," and experiments presented by the Liberty Science Center. During the day the school's schedule is completely turned over to science, math, and technology workshops for children. In addition to the teachers, many parents also volunteer to offer workshops based on knowledge they have gained from their interests, hobbies, or work. This is also the day that parents and teachers lead groups of children in the development and care of the many outside gardens.

Parents as Decisionmakers

The School Site Council, formed in 1987 and composed of parents, teachers, administrators, and a school board member, has established and overseen goals that promote the continued improvement of the school. For instance, after reviewing data on the academic performance of students, the council planned ways to improve programs. The council also provides a forum for discussion of educational issues and sponsors programs for parents and evening family events. Miriam Lally, a parent, values her participation on the council:

> [For] three years I became more involved in the school at large by being a member of the School Site Council and then a co-president for two years. Involvement in this organization clearly gave me a deeper understanding of the school, its procedures, its philosophies, its vision, etc. I derived a sense of satisfaction by being able to contribute to the school whenever we ran a function (Author's Night, Family Read Aloud, Math/Science/Technology Night, etc.) I also learned more about school district happenings since a member of the Board of Education and a district administrator was on the council. Being part of an organization such as this one enables one to gain a much deeper understanding of the school and district than can be learned by reading a school newsletter or community newspaper.

The number and variety of council-sponsored programs show that the work of the council extends benefits to many parents, not just those who are members of the group.

CLASSROOM ACTIVITIES

In addition to schoolwide activities planned to include parents, teachers in the school make connecting their classrooms with parents a priority. The Greenbrook Elementary School faculty and staff are varied in their approaches to teaching but singular in their respect for children and parents. Principal Pat encourages all teachers to approach children and parents in ways that are consistent with the core beliefs of the district and the school but also in ways that feel comfortable and authentic for them as individuals. In this way, teachers on the faculty have developed their own unique styles.

The kindergarten and first-grade teachers invite parents to play important roles in the classroom as volunteers on a regular basis. Parent Carol Desmond describes the kindergarten volunteer program:

> About the second week of school a note comes home to the kindergarten parents asking them whether they would like to volunteer to help out in their child's classroom. Parents can volunteer to work in the classroom either mornings or afternoons once per week or once every other week. This is a true leap of faith on the part of the kindergarten teachers and the principal. They trust in the general abilities of parents enough to say, "You are a welcome addition to my classroom," without even knowing the abilities of the individual parents. The parent volunteer program is so successful that some of the kindergarten teachers have a parent helper all day every day of the week. It is no easy task to plan for, and effectively utilize these parent helpers, but, based on my observations, the kindergarten teachers do this with ease. Parents truly feel they are contributing to the learning process in their own child's class. It is a rewarding experience for parents and appears to be a real win-win program, with the students being the ultimate benefactors.

Another parent, Shari Rothstein, describes her experiences:

> Being involved in my children's classroom is the highlight of each week. As I enter Mrs. Woods' classroom, everything stops as the children greet me. The friendly greeting is always followed by warm hugs. This makes me feel wanted and appreciated.
>
> I come into the classroom to read one-on-one with children, an opportunity they would not have without parent volunteers. I read to the class as a whole or help out on a more involved classroom project. I've even been invited to share my heritage. The students along with the teacher listen intently and ask wonderful questions. I believe my sharing gives them

knowledge they would not receive otherwise, and it teaches tolerance of other cultures.

Teachers in other grades invite parents to help in specific ways for different projects. For instance, primary-grade students are encouraged to write using a unique imaging process. Using a variety of paint techniques, they make a personal portfolio of textured papers, create collages from these papers, and then write stories that emerge from their collages. Parents assist at all points of the process—creating the papers and collages, writing the stories, and publishing the books. During the 2001–2002 school year, Lori Colquist, a second-grade teacher, had a dedicated cadre of parent volunteers working on this project.

Lori Woods, a K–1 grade teacher, invites parents in to be "mystery readers." She values the contributions parents make: "These experiences have greatly enriched the literacy program of all of my classes. . . . I believe that the involvement of parents in the mystery reader program is one of the most successful and meaningful activities that our class engages in. The enthusiasm never fades, in either the children or their parents, and we have readers booked until the last week of school every year!"

Teachers in the upper grades also depend on parents' help for projects. In Thea Dahlberg's fourth-grade class, for example, children create their own animated video with the help of parents. A visitor to the class during the production phase of the project would see parents working with groups of children fully engaged in making their clay figures and background murals, choosing complementary background music, or positioning and repositioning the clay figures—all the while running the video camera to get the animated effect.

Each year, another fourth-grade teacher, Ann Sisko, has her children select and tell Native American stories. Parents work hard as storytelling coaches for small groups of children. The weeks leading up to the festival find the parent coaches working diligently with small groups of children in hallways and in the faculty room, helping them develop skill in the techniques of the oral tradition of storytelling.

Parents see benefits for themselves and for their children from volunteering in their children's classes. A very active parent, Miriam Lally, explains:

> My oldest child completed five years of school at Greenbrook, my second child four years, and my youngest has finished her first year. I have experi-

Sidebar 7.3

Lori Woods, Teacher
Greenbrook Elementary School

Mystery Readers

I hesitate before I make the announcement. Timing is everything, and I know that the children will lose their focus on whatever task they are engaged in as soon as I say the words. Almost everyone is finished. I ring the chime. As each child stops to look and listen, I say, "We have a Mystery Reader today!"

Immediately the room is filled with the buzz of children trying to determine who the mystery reader will be. "I think it's my mom, because she's not working today," says Skyler. "I think I saw your dad in the parking lot," Iris tells Meghan. "My grandma is visiting today and she said she would be a mystery reader," Adam reports. Each child is hoping that it will be his or her mom or dad who will come to our door, sit in our rocker, and share a wonderful story with us.

After a few minutes of conjecture, the children begin to prepare for our visitor. We like our room to be tidy and welcoming so that our reader will know how much their visit means to us. Although I rarely see who does it, there are always two chairs placed beside the rocker by the time the children move to their movie theater seats. One chair is for the child of the Mystery Reader, who will sit beside his or her parent to enjoy the faces of his or her classmates as they savor the story. The other chair is for the occasional siblings who come to listen in.

The call comes from the office. Our reader has arrived. The children sit silently and wait and wait in eager anticipation. I see crossed fingers and hopeful faces. As our visitor approaches, we call out in unison. "MYSTERY READER, PLEASE ENTER!" Without fail, one very excited child jumps up and runs to greet the reader as she or he enters the room. Often many other students follow suit to give our visitor a hug. No matter whose parent it is, we know that we are all going to enjoy the experience that will follow. Once greetings are complete, we settle down to hear some wonderful literature, presented by one of our parents.

A primary goal of our early childhood education is the development of literacy skills in our youngest students. Another priority that I have is to involve parents in our classroom in meaningful and exciting ways. With the Mystery Reader Program, both goals are met. The children are immersed in countless experiences with rich and wonderful literature selections and parents are playing a vital role in the process.

Many parents find ways to embellish the reading. One mom tied a helium balloon to each child's chair while the class was in the music room. When we returned to hear her read *The Balloon Farm*, we felt as if we were actually inside the book. A "penguin mom" arrived one day to read the story of "Tacky" to all of us. We have shared cheese curls during a reading of *Lily's Purple Purse*, because that is Lily's favorite snack. We have manipulated bagel dough after hearing the story of the best bagel maker. One parent brought each child a painter's cap and we all acted out the story of *Caps for Sale*, with her daughter playing the role of the

(continues)

> **Sidebar 7.3** *(continued)*
>
> sleeping salesman and all the other children playing the parts of the mischievous monkeys.
>
> These experiences have greatly enriched the literacy program of all of my classes. One father, who first read to my class four years ago, continues to come each year to share his extraordinary storytelling skills. With his voice, he can change dialects, gender, and age behind the lines (to our continual amusement and delight). Each year I watch the children emulate these magnificent role models when they take their place in the rocking chair to share their own stories with the class. I believe that the involvement of parents in the Mystery Reader Program is one of the most successful and meaningful activities that our class engages in. The enthusiasm never fades, in either the children or their parents, and we have readers booked until the last week of school every year!

enced various levels of parent-class participation as determined by the individual teacher.... My second child's first-grade teacher asked for parent volunteers frequently for such things as in-class baking projects, "mystery reader," and helping children with a book writing project using the "imaging" process. She also invited parents for several special programs such as an "Authors' Tea" and a "Strega Nona Restaurant." I also served as a room parent that year. Her second-grade teacher also asked for parent volunteers to read to the class on special occasions, help with the "imaging" book process, and help at computer time....

Since I have had so much more experience in my daughter's classrooms than my son's, I have felt much more in tune with her education. I have had a clearer understanding of how her first- and second-grade classes were conducted, what the routines were, who the classmates were, how the desks were arranged, etc. I got to know her teachers better and had a good sense of who they were. They certainly gained my trust and confidence in them by seeing them in action—with regard to teaching and how they spoke to the children and handled situations. I cannot say that they necessarily gained my confidence more than my son's teachers; they just did it differently. My son's teachers demonstrated their teaching skills with measurable results as evidenced by papers that came home and by my son's developing reading skills. I never got to see them in action so they remained somewhat of a mystery.

Parent Shari Rothstein says:

> Being in the classroom offers me an opportunity to get to know my children's classmates and to build a relationship with them as well as the teacher.

I also have a sense of security because I get to see what goes on in the classroom. I feel a part of my children's education. It is nice for them to see their parent and teacher as "friends" working together in the interest of their education. They feel proud when I am there. The look on their faces when I walk in the door says it all. I do feel bad for the moms that work full time and don't get this awesome opportunity.

Teachers also understand that parents are children's first teachers and respect the knowledge parents have about their children and how they learn. They invite parents to tell them about the individual children so they can teach them more effectively. They get this personal information in a number of ways. All K–2 parents, for instance, complete a questionnaire to help teachers see the student through the parents' eyes. One K–1 grade teacher, Lori Woods, makes home visits each summer so that she can introduce herself to children and their parents and get to know the children before school even starts: "The home visits give me an important opportunity to receive input from parents regarding their hopes for the coming school year. I leave the families with a letter, requesting them to give me the goals that they have for their child. This input really helps me to formulate the best possible instruction, which will meet not only my goals for the students, but the parents' goals as well."

Fifth-grade teacher, Thea Dahlberg, intentionally uses her parent conferences as a way of building a collaborative relationship with parents. She says: "At first conferences I tell them that I barely know their child and they know the child very well; therefore, I encourage them to tell me about their child. In addition, I ask how they think their child is progressing, what improvements in academics and social skills they'd like to see, and, most important, what they are seeing at home as it relates to school. Using this approach sets up a partnership relationship, and, I believe, lets the parents know that I value their knowledge."

Teachers also listen to parents in other ways. One third-grade teacher, Mona Reisman, told her student teacher, Bonita Williams, this story about how she revised her favorite homework assignment:

> On back-to-school night I began to address book projects. One of the parents had a child in my class a couple of years ago. This particular parent remembered the book projects and immediately gave a dreadful sigh. She knew that the book projects were going to be once a month, which is a lot of work for the students as well as the parents. I knew why this parent sighed so

Sidebar 7.4

Lori Woods, Teacher
Greenbrook Elementary School

Home Visits

One by one, 21 students enter our room. I meet them at the doorway on this first day of school, as I will every day of the school year after today. I know these children already. These are not 21 strangers, but 21 unique and special children who I am already coming to understand. You see, I have visited each of them, at their home, in the weeks prior to the opening of the new school year.

The first bell rings and Adam has headed straight for the computer to try the program we talked about. Alec greets me with a big hug and goes directly to his basket to look at his personalized pillow. Amy stops to tell me that Jack, her new puppy, cried all night. Amber gives me a quick hug and heads right into the room, with barely a wave to her mom. There are no tears, no one hesitates, and each child comes in confidently and with eager anticipation.

I have always done home visits. The teacher I admired most told me, as she retired, "Just do home visits, you'll see why." Because I respected her so much and had already learned so much from her, I set up my first home visits. Nervous and unsure of myself, I rang the bell at the front door of my first family's house. Within minutes I knew why I was making these visits and I knew that they would be a part of my yearly traditions.

When you meet children on their own "turf," they are so much more at ease than if their first encounter is in an unfamiliar classroom, with all the uncertainties that a new school year brings. It helps the children and me so much when we have already spent some time getting to know each other, one to one. At the home visits, I learn about each child's favorite hobbies, sports interests, siblings, favorite toys, and pets. Often children share books that they can read or artwork they are proud of. Many times children take me on a tour of their rooms to show trophies or stuffed animals that are of special importance. Each of these details gives me a clearer window into each child. All of this information helps me to make the instruction I deliver more individualized.

Home visits also give me the opportunity to explain our routines and expectations to the children. I can see their excitement when I explain the role of the line leader in our classroom. When I describe the center time activities that will be available for them, eyes light up and I can sense the enthusiasm for school beginning to grow. I go over each of the items they can expect to find in their baskets, including personalized pillows, and word books to help during writing workshop. On the first day of school, I know that each child will quickly find his or her own basket to explore each of the items that I described.

An important aspect of my home visit is the child's vote for our class mascot. This graphing activity provides a mini math lesson while helping us to select not only our class name but also the endangered species that we will learn about all year as part of our science explorations.

The home visits give me an important opportunity to receive input from parents regarding their hopes for the coming school year. I leave the families

(continues)

> **Sidebar 7.4** *(continued)*
>
> with a letter, requesting them to give me the goals that they have for their child. This input really helps me to formulate the best possible instruction, which will meet not only my goals for the students but the parents' goals, as well.
>
> Before my visits, I send a note to each child to tell them how pleased I am to be their teacher and to tell them that I will be calling to select a date for a home visit, if that is OK with them and their families. A few days later, I call each child. Rarely has my offer to visit been declined, but when it has, there has always been an extenuating circumstance (such as the family is in transition from one house to another). In that case, I always offer to meet with the student and family in our classroom prior to the start of school. That offer has never been declined.
>
> When I first started, the home visits lasted approximately 20 minutes. Now, with all the families I have come to know through sibling students, my visits often last more than an hour. I believe this time is the best investment I can make in my students and their families. The partnership is founded so early. We begin working together before the first school bell rings, and I believe this personal introduction helps to alleviate the anxieties of all those involved: the students, their parents and me. The first day of school is more like a reunion, and a very happy one at that.

deeply because a number of parents over the years had spoken with me about the problems caused by this assignment. I assured the parent that I had heard her cry and changed the book projects to only three times a year. The parent gave a sigh of relief.

Bonita Williams reflected, "It was good to see that the parents have an important voice in the classroom." This story and the important message it conveys will likely stay with her when she begins her own career as a teacher.

FUTURE CHALLENGES FOR GREENBROOK SCHOOL

Even though Greenbrook Elementary School has successfully created an open and inviting climate for parents, children, faculty, and staff, there is still more to do. One way the school gets information from parents about their concerns is through the school district's annual school-climate survey of randomly selected parents. Some Greenbrook parents who responded wished that there were more opportunities for individualized attention for children who were gifted and talented or who had specific learning problems. But most of those who responded were pleased. This response is typical of others: "The pride the teachers and principal have for the school and learning is

> **Sidebar 7.5**
>
> *Thea Dahlberg, Teacher*
> *Greenbrook Elementary School*
>
> ## Parent Conferences
>
> Everything I do at conferences I learned from my principal, Patricia Holliday. I sit at a cluster of desks with the parents. I try to encourage them to sit next to me so that it is easier to regard work together. At first conferences I tell them that I barely know their child and they know the child very well; therefore, I encourage them to tell me about their child. In addition, I ask how they think their child is progressing, what improvements in academics and social skills they'd like to see, and, most important, what they are seeing at home as it relates to school. Using this approach sets up a partnership relationship and, I believe, lets the parents know that I value their knowledge.
>
> When a parent comes to me with a concern, I take a deep breath. In this deep breath is the unvoiced message (in other words, a reminder to myself) that this person loves his or her child deeply and that everything she or he is going to say comes from that love and concern. Such a reminder prevents defensiveness. I don't think it's very effective for teachers to tell parents that their concerns are groundless. These concerns come from somewhere and need to be addressed in some way. For example, if a parent is concerned about our spelling words not being challenging enough, I make sure I validate that parent's worry about the importance of developing good spelling habits and a viable writing vocabulary. I then explain the many ways we develop spelling skills and how a greater writing vocabulary comes through the study of content.
>
> The most important thing I have learned is that if I communicate my affection and concern for their children, parents are far more willing to listen and to trust my decisions.

passed on to the students. My daughter loves this school and all the teachers. I am grateful for all everyone does for her."

Of course, in the rough-and-tumble of running any school, problems will arise—parents have concerns, raise questions, get angry. What is different about Greenbrook, however, is that concerns are taken seriously and quickly attended to. Principal Pat says her number-one rule is "Don't ignore it!" Parents are heard, and issues are resolved. She also knows that trust is fragile. She continuously reminds her staff, "We're only as good as that last interaction. Everything can get wiped out by one mistake. We have to continuously be proactive."

Each year Greenbrook Elementary School, like many other schools across the nation, opens its doors to an increasing number of families from different cultures. Greenbrook, which had few families of difference a decade ago, now has almost 25 percent of its population with roots in China, India, Mexico, and the African American community. Parents from these cultures are not well represented in some of the parent-involvement activities, such as the Kindergarten Journal Writing Project, and, for the most part, they don't take on leadership roles in the school's parent organizations. However, many of them do attend events where their children are performing, attend workshops for parents, and will, if personally asked, serve on committees or help with short-term projects.

Principal Pat and her staff are committed to learning how to make these parents feel welcome. For instance, many teachers invite parents into their classrooms to teach students about their classmates' cultures. And in the process, the teachers gain knowledge they need but didn't have. Pat says the key is to reach out in a variety of ways to all parents, but especially to those who may feel uncomfortable in this very mainstream American, middle-class school. She notes that the personal one-on-one invitation is the most important. Thus, she and the teachers are making an attempt to make more personal contacts with parents.

Cultural differences may also influence participation. A recent student teacher at the school of Pakistani descent (and a mother of a kindergarten child herself) has suggested that participation in programs, such as the Kindergarten Journal Writing project, may run counter to the cultural expectations of East Asian mothers who have a large network of extended family with whom to discuss child-rearing problems. When parents come from another culture and are educated in other countries in schools where the teacher is the unquestioned authority, they may have different ways of thinking about the role of parents in schools. A challenge for the Greenbrook faculty and the white, middle-class, stay-at-home mothers who make up the majority of the parent leadership in the school is to continue to seek new ways to help parents from different cultures to feel comfortable becoming more involved with the school.

Greenbrook also faces another challenge: The school faces the unsettling task of dissolving and rebuilding its community as a result of a complete redistricting of the current school population. Because of the rapid growth in the area, a new elementary school opens in 2002. Only 40 percent of Greenbrook's current school community will remain after the redistricting; 60 percent of the school population will be new to the school in September of 2002. Creating a new community will be a major challenge for the faculty and the parents. As one parent says, "Greenbrook has been my community as well as my child's. I have made friends here. I know everyone. I belong. It will be difficult to go to a new school where I have to start anew." Parent Miriam Lally feels the loss that will come along with the changes:

> Having been at Greenbrook for seven years, I have developed relationships with faculty and students and feel like I really know the school and am part of a special community. Redistricting will affect my family. My middle child may stay here through fifth grade. My youngest will definitely be relocated to Brunswick Acres School. We'll need to adapt to a new environment and develop new relationships. That wouldn't be so bad if we didn't have to leave the relationships that both my children and I have formed. I can't help but feel the loss already. It will be strange for us to go from being well-known members of a small community to strangers in a new place.

While some parents worry about finding community in a different school, those who will still be at Greenbrook will have to work hard to recreate the community feeling that exists at the school today. As they welcome the newcomers and integrate them into the culture of strong parent participation that currently exists, the culture will inevitably change. Yet Greenbrook School seems well prepared to deal with this significant challenge. With such capable leadership, outstanding teachers and staff, a tradition of active parent participation, an ethic of caring and inclusion, and an infrastructure of parent participation activities, this school community has a strong foundation on which to build. In a multitude of ways, Greenbrook seems ready to meet this challenge and others as yet unknown.

8

CREATING A COMMUNITY OF DIFFERENCE: CROSSROADS MIDDLE SCHOOL

> *[When] adults take the time to wrestle with the difficulties of these issues [racism], it communicates in a powerful and meaningful way to the children that we care enough about them to do this.*
>
> —Marjorie Moore, parent,
> Crossroads Middle School, e-mail message

Talking about race is not easy—especially in a mixed-race group. In fact, conversations about race are so difficult that most people avoid them. This was not the case at Crossroads School, a small middle school of choice in the upper West Side of Manhattan. During a parent association meeting of 30 participants devoted to a discussion about how their children responded to reading, the issue of racism arose. And it wasn't avoided. Some African American parents took the risk of exposing their pain openly and honestly to this group of Latino, white, and other black parents. Following the meeting, an African American parent, Carolyn Jordan, shared the following message online via the Parent Association Steering Committee Yahoo!Groups discussion list subscribed to by 26 other parents and teachers:

> I think that we had a good meeting last evening. However, I do want to say that I hope that we didn't make anyone too uncomfortable with our open and honest talk. It certainly was not intended to make anyone uncomfortable and/or feel guilty. The problems with these kind of issues

---

Sidebar 8.1

Carolyn Jordan, Parent
Crossroads School

A Parent's View

Black History Month Must Be Celebrated in Our Schools
I believed strongly that the school should celebrate Black History Month. Ann, the director of the school, believed that the study of African American history should not have to be singled out as separate from the study of history—that it should be part of the larger course of study. I agree in principle; however, the accomplishments of Black people are not made part of the teaching of American history, unless it's the Civil War that's being discussed. A Black child's self-esteem will not be increased if all you talk about is slavery. There needs to be more positive truth about Black achievement discussed in classrooms. When discussing slavery, they must talk also about the effect this had and has on a people. Then, in order to build or rebuild self-esteem, you must talk about how far Black people have come in spite of the horrific crime of slavery. Because the positive about Black people is not part of the normal curriculum and only focused on during Black History Month, it is necessary for it to be celebrated. I feel very passionate about this. The African American experience in this country is very different from the experience of any other people. We are the only people brought to this country and stripped of anything that would give us any connection to the place that we come from, Africa. Yet I don't know of *any* Africans in my family. We are the only people in America that have no country to identify with, making for an experience that cannot be compared to any other people who have come here from other countries. It is very necessary for Black children to understand the contributions made by African Americans to the development of this country.

---

and the reason why we do not deal with the racism problem as a society is because it makes people feel edgy.... I hope you understand that as victims one does not have the choice of not dealing with the problem. Although I realize I said a lot, I felt there was reluctance to respond by most. Probably because everyone is afraid of offending someone.[1]

Harriet Bograd, a white parent, responded:

I didn't view it as a reluctance to respond. I thought we were going around the room each telling about our kids' response to reading and that I should wait to respond until it got around to my turn. Also I was enjoying listening—it was great to hear others speak so passionately.... I was really grateful that you, Carolyn, and many others shared your thoughts so openly. And

> **Sidebar 8.2**
>
> *Ann Wiener, Director*
> *Crossroads School*
> *The Director's View*
>
> ### Multicultural Curriculum: What Is It? How Do We Do It?
>
> Staff and parents have been talking the last few years about multicultural education and struggling in particular with what to do about Black History Month and Women's History Month. Like all of us, I have seen the importance of including race and gender in our curricula. However, I have been uncomfortable with doing it at a specified time.
>
> This summer I read *Invisible Privilege* by Paula Rothenberg, professor of philosophy and women's studies at William Patterson University of New Jersey and director of the New Jersey Project on Inclusive Scholarship. She explained my discomfort as she wrote "teaching and talking about 'women' one month and 'African Americans' another constructs the majority of people in our societies as 'minorities' and 'other.' It implies they have not earned the right to a place in the 'real' curriculum and can enter only by means of a kind of curricular affirmative action that sets aside certain places for them."
>
> Paula Rothenberg also described what we try to do at Crossroads: "teach a curriculum that acknowledges the race, class, and gender differences that have shaped every aspect of the world we live in today ... teach our students to identify their impact and to understand that eliminating their pernicious effects is in everyone's interests." As she says, "Owning the past can only enhance the future. Failing to acknowledge our past leaves the present confused and confusing. Naming injustice for the sole purpose of assigning blame will divide us; identifying injustice in order to address it has the potential to move us toward a society that values both diversity and community."
>
> Can we all—parents, students, staff—explore multiculturalism this year? How do we define it? What is the school doing through curriculum, advisory, etc.? What do the parents see as their needs as they raise children in a pluralistic society? What can we all do better?

I love the fact that our meetings are free to go off on important issues like this—at the testing meeting, we all talked about reading, and at the reading meeting, there was lots of talk about race relations.[2]

And the school director, Ann Wiener, also white, added:

> Thank you, Carolyn, and all those at Thursday night's meeting who have raised the essential and urgent issue of our time. Unfortunately W. E. B. DuBois is still right when he wrote so many years ago that "race is the issue of our century."[3]

In another message, Ann said:

> I did not feel uncomfortable or guilty last night. I was quiet because I was wrestling with what you were saying and revisiting my thoughts about the issue. Please know that I take very seriously being head of a school diverse racially, ethnically, culturally, and in ability. I am well aware that all whites no matter how much reading, experiencing, and reflecting, etc. participate in the "presumption of whiteness" and must constantly remind ourselves of this. I appreciated the open and honest discussion last night and did not feel a reluctance to respond by most.[4]

What started as a meeting devoted to issues of reading turned into a powerful dialogue about race that led to a discussion about how to adequately address racism in the curriculum, how to make Black History Month and Martin Luther King's birthday celebration more meaningful, and how to teach about the powerful role of the church in African American history in a multicultural community. Ann asked parents to help her and the Crossroads staff to rethink how the school addresses racism. She sent the following messages to the discussion list: "Perhaps racism should be the subject of one of our PA meetings this year. You also make me rethink Martin Luther King Day and Black History month. I hope others respond with suggestions of how we can use the next month at school to help students reflect about these issues. . . . I shared last night's discussion briefly with the staff and we are thinking about what we can do for Black History Month. All suggestions are welcome and thanks for yours"[5] and "Can we respond to Marjorie's [a parent] question 'What would two days of a program devoted to providing an experiential journey about racism look [like]? I would also add, 'How can we extend it to racism as it affects others: Latinos, the poor, Asians?'"[6]

The online discussion group brainstormed ways to address racism in a meaningful way and developed a parent-planned and initiated Martin Luther King celebration that included the singing of "Lift Every Voice and Sing," the Black National Anthem, a read-in of favorite selections related to African American history by parents, students, and staff, and storytelling by African American storytellers. In addition, parents developed a list of resources for Martin Luther King celebration events throughout New York City that were distributed to parents and staff.

This story tells a lot about the underlying beliefs that provide the foundation for parent/school relationships at Crossroads. Here parent voices are respected. Here there is a realization of the importance of building an open and inclusive community of parents. Here parents trust each other and the staff enough to raise difficult issues. Here there are many ways for parents to be involved and heard, and there are many different ways to communicate with each other.

CROSSROADS SCHOOL: A COMMUNITY OF DIFFERENCES

Founded in 1990, Crossroads School was created out of the hearts and minds of Ann Wiener and two teacher colleagues, Kathe Jervis and Clyde "Buck" Creamer. Frustrated by the monolithic bureaucratic schools that made up much of the urban educational landscape, Ann and her colleagues were convinced that children could be best served in small personalized places—places where each child and her or his family would be known, respected, and invited into an exciting learning environment. They believed that urban youngsters deserved a curriculum that spoke to their experiences and that differences enrich their learning. During the late 1980s and early 1990s, when the media reported that urban schools were failing to educate the children they served, cities like New York began to listen to reformers, such as Ted Sizer and the Coalition of Essential Schools, as they looked for alternatives to their failing systems.[7] One approach taken by New York City districts was to create small schools of choice. Taking advantage of this interest in creating alternatives, Ann and her colleagues presented a proposal to District 3 in Manhattan for a small middle school of choice—parents would apply to send their children to the school and the school would choose students whom they could best serve. They found a possible site, the fifth floor of a large 100-year-old elementary school building on the Upper West Side of Manhattan. When Ann visited the big open space, which had been condemned because of roof leaks, she found beautiful Ionic columns holding up the ceiling and home to pigeons and their nests. Recruiting a staff and planning the school was her first challenge. Getting the roof repaired and the classrooms prepared was her next. The school opened in September 1990 with 60 students and 4 staff members. Since then it has gained students and staff and has built a reputation among parents as a place to be trusted with the education of their children.

The school became part of a bustling, diverse, and vibrant community. This area is home to Latino immigrants from a variety of Spanish-speaking countries; liberal-minded, urbane intellectuals—black and white; artists, students, and others, both well off and poor. Many mixed racial and ethnic families choose to live here, comfortably blending into its diverse population. The parents who choose to send their children to Crossroads are representative of all these groups. Additionally, some parents from Harlem, the Bronx, and Brooklyn also choose this school for their children.

In 2001–2002 the school enrolled 209 youngsters in the sixth, seventh, and eighth grades (an increase of about 50 from the previous year due to decisions made by the district office—decisions that placed significant burdens on a school devoted to making sure that each child is known as an individual by all of the teachers and staff): 5 percent white, 3 percent Asian, 48.5 percent Latino (many first-generation students from the Caribbean), and 42 percent African American. These percentages are somewhat representative of the larger urban environment where the school is located, but the statistics do not include the many mixed-race children who attend the school. Because the school was a pioneer in the inclusion effort in the mid-1990s, now more than 20 percent of the students are identified as needing special-education services.

The staff, on the other hand, before 2000–2001 had not been as diverse as the student population. Thanks to a special effort to increase faculty diversity, 5 of the 13 teachers now are persons of color. There are 9 other staff members, some part time, including a speech and language teacher, English as a Second Language (ESL) teacher, and a guidance counselor, as well as teaching fellows, student teachers, and office staff. Other people who work in the school do represent both the Latino and African American communities. For instance, the office manager, who has served as an advisor for a student advisory group in past years, is a major asset: She not only translates written documents from English to Spanish, but she also explains the rationale for the school's practices to the Latino community of which she is a member.

PARENTS CHOOSE CROSSROADS FOR MANY DIFFERENT REASONS

In School District 3 in Manhattan, as in some other New York City districts, parents apply for admission to middle and high schools of their choice. When their children are in the fifth grade, parents are given a book that describes each of the schools in the district. Then, with the help of elementary-school guidance counselors, parents visit schools that interest them and attend school fairs. In December they indicate their first four choices on an application form. Applications are then sent to each parent's first-choice school and then down the line through the four choices until the child is accepted by one of the four schools. The Crossroads staff also interviews prospective students, gives them a simple math evaluation, and asks them to complete a writing sample. Using multiple criteria and with the goal of keeping the school as diverse as possible in mind, the full staff at Crossroads make the admission decisions. Because parents are not notified about the schools that have accepted their children until May, the period between December and May is a time of much stress as parents worry whether their children will be admitted to the schools they prefer.

Because parents choose to send their children to Crossroads, it would seem that they would support what the school is doing, that is, parents chose this school because they understand and agree with its stated philosophy. This is not necessarily the case for many parents, however. Some parents, because they have the know-how, time and resources, do extensive research to choose the schools best suited for their children. Other parents, burdened by the day-to-day struggles of making a living and learning how to adjust to a new culture and language, depend instead on the recommendations of family and friends or select the closest and most convenient school.

Parents give different explanations for choosing Crossroads. One parent, for example, said, "Crossroads would provide a familiar place for [my son] since it was in the same building as his elementary school. He could still see some of the teachers he knew from elementary school." Another parent, though, considered the match between Crossroads' nontraditional teaching and the learning styles of her children: "I read an article in either the *New Yorker* or *Time* magazine. It talked about the best schools in New York City. Crossroads was mentioned as one of them. It was described as nontraditional. While that isn't my personal way of doing things, I thought it would be good for my son since he's nontraditional." And then there are parents who send their children here because the school focuses on children's social and emotional growth as well as their academic development: "We were looking for a school that was more humanistic than those that are highly academic and competitive. They [teachers at Crossroads] use a discovery form of teaching that allows students to discover the subject." Other parents decided on Crossroads after visiting the school and sitting in on a class, as one parent interviewed for this chapter explains: "During the visit my daughter attended a humanities class and fell in love with humanities and with the teacher." Her daughter was so completely taken with the teacher that she said, "This is where I'm going to school!"

Parents, of course, bring their own understandings and create their own visions of what Crossroads is and what it will be like for their children. Given so many different perspectives, some parents will misunderstand or be unhappy with some of the school's practices. Even those parents who choose the school because they agree with the school's stated philosophy may not actually understand what that philosophy will look like in practice. As another parent who was interviewed noted, "All parents are aware of the school's philosophy on some level. But it is different when you see it in practice. Some parents use their own school experience as a guide for what they think should go on in schools. So if they got a lot of homework and carried lots of textbooks back and forth to school each day, they think that's what makes a good school. That's the way they measure whether or not their child is getting a proper education. At Crossroads that doesn't necessarily happen and they get nervous."

Crossroads works hard to help parents understand and support the school's mission; this effort is always ongoing. One former Parent Association president pointed out that "We continue to struggle with how to better communicate to parents the school's philosophy and teaching style in order to allay their concerns and so they will be better equipped to partner with the school so that their child can get the best education possible." Despite what everyone does, he notes that the task can be difficult: "While as PA president it was my responsibility to help parents to trust the process, it's hard to trust the process when your kid is not doing well. If things aren't working out for your youngster, you can't hit the rewind button and do it over."

CROSSROADS SCHOOL IS PROGRESSIVE— AND TRADITIONAL

The heart of the school's educational program is built on a foundation of progressive thought about how children learn and teachers should teach. The school's philosophy is presented in the school's handbook:

> Through a carefully designed, relevant curriculum and a nurturing environment, we develop the civil and social skills necessary to function productively and ethically in the world. We aim to create a community of lifelong learners.
>
> We believe children learn best in a small setting where they can be known. We want our students to become critical thinkers, able to analyze and impact the world around them.

To achieve these goals, Crossroads features a student-advisory system, block scheduling, multi-aged grouping, cooperative learning, and community service. We do not track our students. Students are assessed in a variety of ways: student self-evaluations, narrative teacher reports, portfolio assessment, as well as state and city-wide tests. Our major goal is knowing each student well as an individual learner so we can help all students reach their potential.[8]

The Crossroads program builds on many of the 10 principles of the Coalition of Essential Schools—small classes, personalized instruction, a curriculum that strives for depth rather than coverage, teachers acting as coaches rather than as sages, and the emphasis on building communities respectful of differences. In addition, its curriculum reflects the belief of Emily Style, a well-known feminist scholar, who argues that the curriculum should provide both a mirror in—a way for the youngsters to see their own experiences—as well as a window out into the larger world.[9] The Crossroads curriculum, based in a pedagogy of social justice, reflects the history and experience of the urban African American, Latino, Asian, and white youngsters it serves.

The staff also strongly values the importance of the social-emotional development of their students, adolescents trying to figure out who they are and how they fit in the world. To this end, the school pays a great deal of attention to community building. The first five days of school are structured so that students can get to know each other and the staff. The school handbook explains:

> Students learn best when they feel a part of a school and familiar with staff, their peers, and school expectations. Therefore, we design the first 5 days of school in a special way. The first three days are half days. Each student only attends 1 of the first 2 days.... With half of the school attending, students and staff get a chance to know each other and school expectations in small groups. All students attend the 3rd day for more orientation activities. On the 4th day, students spend an entire day with their advisory.... The 5th day is our first all-school trip. This educational and interesting event prepares students for a year of active learning. Classes begin on the 6th day of school; by this time students feel comfortable and ready to learn.[10]

There is other evidence of the school's strong commitment to the social and emotional growth of students. A three-day overnight camping trip to Camp Mariah, a Fresh Air Fund camp in Fishkill, New York, focuses on community-building activities; ice skating and roller skating trips to Central Park encourage children to take risks. Because the faculty recognizes that young adolescents don't all develop at the same rate, sixth, seventh, and eighth grade students work together in multiage groups in most academic classes.

On the other hand, expectations about how students should behave have become more traditional over the years. Early on, for instance, Ana Chanlatte, office manager, student advisor, and parent of two children who have previously attended Crossroads, expressed concerns about the lack of a consistent approach to issues of discipline: "At Crossroads, there is 'invisible' discipline. Rules exist, but they are not always followed."[11] Her concerns and the different perspectives of staff members—some less demanding than others—led to staff discussions and finally in 1994 to a Code of Conduct that both students and their parents are expected to sign.[12] The handbook explains the basis for this code:

> Crossroads' rules and regulations reflect a belief that in order for students to learn and teachers to teach Crossroads must be a safe and productive environment for everyone. Therefore, we focus on helping students create and implement appropriate ways for dealing with conflicts that do not involve violence and threatening language. We also strive to help students develop personal responsibility, independence, and self-control so they can become productive and successful adults. We feel that all members of our community have a responsibility to one another and to the community as a whole.
>
> Because adolescence is a time when many kids test their boundaries, behavior issues and conflicts will arise. With that in mind the Crossroads staff has developed the Crossroads Code of Conduct to outline specific behavior expectations and disciplinary procedures.[13]

It took a number of years for the entire staff to consistently enforce the new code. As Ana said in 1996, "Until we really get everyone together to see the same things, we will still have problems with discipline. We still have different values." However, at this point there seems to be a consensus about the importance of enforcing the code. One of the veteran teachers has even been freed up to act as the "Dean of Discipline" to make sure that the code is enforced consistently and fairly. Parents recognize and appreciate this and say that one of the things they like about Crossroads is its "loving firmness." One parent explains why:

There is a belief in firmness. Sometimes progressive schools can be loose. Here there are high expectations for appropriate behavior and consequences for inappropriate behavior. For example, the Peanut Butter Lunch. (Don't you just love that name?) It captures the "loving firmness" of the school. When children act inappropriately they are assigned to PBL where they lose their privilege of going off campus for lunch. It reflects the approach described by Ron Taffel, a family therapist, who spoke to a Parent Association meeting at the school. He says we need to provide a "caring envelope of firmness and love."[14]

Another parent observed, "While this school provides a liberal school experience, it is not like a school without walls. There are traditional ways of doing things. For instance, teachers are not called by their first names. Traditional respect is required. I know that if you give my son too much freedom, he'll take advantage of it. So I was happy that they provided structure for him."

CROSSROADS VALUES PARENTS

Crossroads is a collaborative community including staff, parents, and students. The parents, children, and staff come from various ethnic, cultural, economic, and religious backgrounds. Crossroads staff works together with families and the community to achieve academic excellence and social competence for all of our students.

—Crossroads School handbook

Everyone at Crossroads works hard to bring these words to life. There are myriad ways (described later) parents come into the school community as participants, volunteers, learners, and leaders. But most important, the school takes seriously the notion of partnering with parents in the education of their children. Central to this commitment is the school's student advisory program. In many schools, because they still adhere to the traditional view of parent/school relationships, tensions between parents and teachers may develop—parents advocate for the needs of their own children while teachers feel bound to a commitment to all children. The student advisory program changes the nature of this tension at Crossroads. Because teachers and other staff members in their roles as advisors are expected to work hand in hand with parents to meet the needs of individual children, they, in fact, do advocate for individual children in some of the same ways a parent might. Parents appreciate the support; as one parent said, "My son's advisor was a wise choice. We stayed in close contact. I felt like I had a partner. We worked closely."

Each student is assigned to an advisory group—a group of 10 to 12 students mentored by a staff member.[15] Advisors act as guidance counselors, mediators, advocates, and parent liaisons. They are responsible for guiding each student in their group through the social, emotional, and academic turbulence of adolescence. All parents know how to contact their child's advisor at school and at home—they have their child's advisor's home telephone number. Parents are surprised because they are not used to such easy access. One parent said, "What blew me away was that teachers gave us their home phone numbers. I thought this was crazy but courageous." Enacting the school's foundational belief that parents and teachers must work together means that they have continuous contact about the child. Teachers know when something happens in the family that may impact the child, such as the birth of a new child, the loss of an apartment, or the separation of parents. And parents know when their child does something wonderful, when they excel or make a breakthrough as well as when they have trouble with a class. In addition to regular telephone contact, parents, advisors, and students meet twice a year for student-led conferences about the children's learning.

Advisors take on many different roles: academic coach, liaison with other teachers, counselor, disciplinarian, and, most important, they are the primary contact with parents. Chanlatte has been an advisor for a number of years (although due to the increase of students, her office work prevents her from taking on this responsibility this year.) She says she needed to maintain a good relationship with both students and parents. She would say to her students, "My responsibility is not only to you. I have a responsibility to your parents, as well. . . . We are not only here for the kids, we're here for the family." As a part of the Latino community, she was well placed to explain school programs to her students' parents. When parents of her advisees weren't sure that they wanted to allow their daughters to attend the annual overnight camping trip, she was able to tell them about her own struggles with this very same issue. Because she has gone on these trips, she was able to reassure them that the staff takes very good care of the children when they are away.

The advisory program's primary goal is to create an educational program that meets the needs of each individual child, but the program also fosters the development of a trusting and respectful relationship between parents and the school's staff—and it shows in many ways. Parents choose to send younger siblings to the school and tell friends to send their children there. Some continue to keep children at Crossroads even when they move out of the neighborhood; as a result many children have long subway trips from the Bronx and Brooklyn. When it comes time to choose high schools for their children, parents seek out and listen to the advice of Crossroads teachers and advisors.

Parents' personal interactions with the school's teachers, principal, and their child's advisor are convincing evidence that Crossroads cares about their children. For instance, one parent said, "What I learned is that the staff has an investment in the children over the long haul—they are not interested in the short term—today's grades—but rather to commitment to learning in general . . . they discover what type of learner your child is and deal with each child where they are." Another parent commented, "Not all parents can articulate the educational vision, but they know that their kids are loved and supported, that their kids like coming to school while they didn't like going to school in their previous schools. They know that they, the parents, are respected." Another parent's comment shows the trusting relationship that can develop between an advisor and a parent: "My daughter needs other strong people in her life to advise her. My husband and I aren't the only people who should be guiding her."

Of course, other parents have less positive views. If a child is not doing well in school and the relationship between the advisor and the parent is troubled, trust can be tentative at best or missing altogether. Parents who have low social status and those who have had bad personal experiences with schools come to this school as they would to any school: suspicious of its motives. The school staff must make enormous efforts to break down that suspicion and to build trust.[16] One parent reflecting on trust suggested, "If there is a core trust that the school works in the best interests of their child, then the parent is likely to trust school decisions, or at least not attribute other motives to decisions. If there is not this core trust in the school, then some parents are prone to look at other possibilities, such as racial bias." And some tensions have emerged between parents and advisors, as a later section in this chapter shows.

Creating a community of difference where all parents, teachers, and students are respected and trusted is no easy task—but that is the goal of Crossroads School. Jianzhong Xu, an educational researcher who studied Crossroads, pointed out it is even harder to do so with parents who bring histories of mistrust of schools to the table: "a school faces more challenges in its effort to reach out to families from diverse backgrounds than to their children."[17]

CROSSROADS AND THE COMMUNITY

The "community" for any school in New York City can be defined in several ways. First, there is the immediate community that surrounds the school where many—but not all—of the families live. But "community" might also mean the area referred to as the Upper West Side, which is home to Columbia University; Bank Street College; a variety of shops, theaters, restaurants, food emporiums; pre–World War II apartment buildings and tenements; and new high-rise apartments. Or "community" might include all of Manhattan, since Crossroads takes advantage of all its resources: Central Park's ice skating rink, museums, and other cultural venues. Whether these are just a walk away or require a subway ride, they are an integral part of the school's program.

Crossroads reaches out to the wider community. Each spring the school, with the help of parents, staff, and students, finds summer placements in camps, schools, and community programs for each student. Thanks to community groups, private schools, businesses, and state and local agencies, Crossroads has successfully placed most of its students in exciting summer programs both in the city and farther away in country settings.

The school is also mindful of its extended local community. When people were roused to anger by the killing without cause of Amadou Diallo by New York City policemen, Crossroads students, studying the Civil Rights era of the 1960s, planned and staged a peaceful demonstration in the neighborhood. Students, parents, and staff also joined the protest. Because Crossroads School is committed to social justice, and because many of the students are poor and come from disempowered groups, the faculty helped students take thoughtful action on their own. The school teaches by word and by example.

THE SCHOOL DIRECTOR AND
TEACHERS SHARE LEADERSHIP

Most educational scholars today acknowledge what common sense has told us for years: The most important person in the school is the principal. The principal sets the tone, carries the vision, and, as noted educational scholar on leadership Thomas Sergiovanni observes, "helps the amoeba to cross the road."[18] Crossroads School is fortunate to have Ann Wiener, a skilled leader, who understands the importance of her role.

The School Director Leads with Vision

Although parents had many different reasons for enrolling their children in Crossroads, the parents interviewed speak with one voice about director Ann Wiener. "I fell in love with her. After she interviewed my child, I met with her privately. We spoke for hours and found out that we had lots in common." Another said, "I was very impressed by Ann. She was so open. It was apparent that she really cared about children." Others commented: "After attending an open house, my husband came home and said 'I just met the most wonderful principal. We have to visit the school,'" and "I've gotten to know Ann quite well. She has helped me define my vision of education. She really listens to parents and knows the children well."

Ann Wiener clearly has captured the trust, respect, and admiration of this set of parents. How does she do it? Ann describes the vision to which she is committed and uses to guide her daily actions and interactions: "The director's most important responsibilities are to articulate a vision that is open enough for staff, students, and parents to become part of, and flexible enough to be modified without compromise; to be strong but not authoritarian; to be a person parents, staff, and students can trust to be fair."[19] She explains why she developed this vision:

> My vision is strongly rooted in my experience. Because I still feel at times like an outsider and value the power and richness of working with others, I want Crossroads to be a place where everyone—staff, students, parents—can belong and grow; an inclusive, not exclusive, community.
>
> Because I know the pain of being seen through a single lens, I want there to be many ways to be successful and acknowledged at Crossroads.
>
> Because I am a mother of African American and Caucasian children, I know we are a nation of diversity; because I have benefited from leaving the confines of white privilege, I want students, parents, and staff to learn from working with others of varying cultures, races, religions, and abilities.
>
> Because I know education must belong to the individual, I want the Crossroads curriculum to be rich enough, varied enough, meaningful enough that all students can connect to it and build their own knowledge.

> Because I know that I learned best when I related to a teacher—especially as an adolescent—I look for a diverse group of open, interesting adults who value working together on behalf of children.[20]

Ann makes her vision come alive in many ways, big and small, in her interactions with parents. For instance, parents get priority in her day. When a parent comes in or calls wanting to discuss something, Ann sets other things aside to attend to their concerns. When parents send e-mails either to the Parent Association Steering Committee or to the School Leadership Team online discussion group, she responds quickly, acknowledging in a thoughtful way what the person said. She recognizes that it is important to hear from parents: "Parents know so much about their children—they must be listened to." She realizes that every contact she makes with parents has the potential to strengthen their trust or to break it down.

Ann knows her role as leader and "moral authority" of the school can be critical: "My leadership . . . is particularly important when the issues involve race, class, and gender. At these times the leader must be the moral authority, the one who makes sure that the circle is expansive and articulated and there are structures in place to deal with these difficult issues, so there can be conversation in many different constellations. Leadership is about providing a wide variety of entry points." Tackling these hard issues can happen only in a school where parents, teachers, and the director have developed a strong sense of trust in one another. The ongoing debate over the role of Black History Month celebrations, described elsewhere in this chapter, is an example of this school's ability to confront different perspectives in respectful and constructive ways. One parent put it this way: "This is a pretty dynamic group of parents. They don't bite their tongues. They're not afraid to speak up. Ann is receptive. She listens and hears what parents are saying. She does not get defensive and thinks about what parents say."

Teachers Are Leaders, Too

Teachers choose to teach in this school because they are committed to its philosophy of social justice and progressive pedagogy. They have an active voice in all decisions related to the school program. They work long and hard. The teachers attend many nightly and after-school meetings. They participate in the fall parent/teacher retreat on a weekend in September and take their students on an extended overnight camping trip in October. They do all these things because they are committed to the school's core values and beliefs. They understand that, in order to educate children well, teachers must attend not only to children's academic needs but also to their emotional and social needs. The work is exhausting.

Parents speak highly of the school's advisee program. While a few parents do worry that perhaps the teacher takes on too much responsibility for their children, thus leaving parents out of the loop, most parents interviewed spoke glowingly of the partnership that exists between them and their child's advisor. The advisor/advisee relationship, although valued by teachers and parents alike, create some tensions. As one teacher explains:

> This is a very messy, nonsystematic relationship. Sometimes we step on parents' toes. And sometimes they step on ours. This is a very taxing job. We give our students and advisees our home telephone numbers. I can spend hours on the phone or answering e-mails. However, sometimes some parents don't understand that they are imposing upon our personal time and space. . . . Some parents are demanding. Others, while demanding, do it in a way that is more respectful, making me more willing to give the extra time. They recognize that we lead very busy lives.

Although this teacher does not want to jeopardize the strong bond that the school builds with families or the commitment teachers have for each individual student, she wants parents to respect her private life.

Teachers play a major role in the development of the progressive curriculum and teaching practices found at the school. Their leadership in areas of curriculum, however, sometimes puts them at odds with some parents. As one teacher said, "Sometimes parents aren't well enough informed about the progressive nature of our school and don't fully understand or agree with our program." Parents who think good schools give lots of homework out of textbooks sometimes don't understand why their children aren't required to do this type of work. As part of their professional responsibility, teachers make presentations about their progressive practices at Parent Association meetings. Unfortunately, not all parents attend these meetings.

Issues such as these also are addressed at the school's fall retreat for parents and teachers, but teachers have left some retreats feeling that parents were criticizing them without also acknowledging their hard work and amount of extra time they put into their jobs. Although Ann has many ongoing conversations with parents about these and other issues, such conversations between teachers and parents are much less common. But it may be difficult for teachers to find time to add something else to the heavy load they already carry. The bottom line is that teaching at Crossroads does require great dedication and commitment; as one teacher says, "All in all, this job is extremely rewarding. I meet wonderful people, have deep, inquiring relationships with them. Teaching is an infinitely messy and interesting job."

CROSSROADS PARENTS AS LEADERS AND PARTICIPANTS

The role of parent leaders in schools is much less studied and understood than that of others as leaders. However, at Crossroads, as in many other schools, a core group of parents emerges each year to take on leadership roles both in the Parent Association and on the School Leadership Team. These parent leaders assume responsibility for nurturing an active parent community. Throughout the year Crossroads schedules events that help bring parents into the school community. The winter Festival of Lights celebrates all of the winter traditions of different cultures represented in the school: Christmas, Hanukah, and Kwanzaa. Poetry, music, and dance presentations delight the packed house of parents and families. The Basketball Bonanza in February finds parents, teachers, staff, and students participating in a variety of basketball-related events. Parents attend Parents Night Out in April and the Arts Festival in May. The June events include the All-School Picnic, the Eighth-Grade Prom, and Graduation. Central to the success all of these activities is the vibrant Parent Association.

Leadership Opportunities Open to All

Many times parent leadership groups are closed cliques. This is not the case at Crossroads. Before students even begin at Crossroads, newly accepted students and their families are invited to join current students, their parents, and school staff members at the All-School Picnic in June. Newcomers thus meet and begin to form relationships with others. The picnic, held in Riverside Park, allows new parents to share stories with other new parents and to confirm, one more time, that they have made the right choice for their children. One parent explained how the picnic is important:

> In June of the year before your child enters the school there is a picnic that all new children and their parents are invited to. It is a time to get to meet

> **Sidebar 8.3**
>
> ## Crossroads School Calendar
>
> *2001–2002*
>
> | | |
> |---|---|
> | Every month | School Leadership Team—staff and parents |
> | | Parent Association Steering Committee Meetings—open to all parents |
> | | Parent Association Meetings |
> | September | Curriculum Night |
> | | Annual Parent/Staff Retreat |
> | October | High School Meeting, 8th-Grade Students and Parents |
> | | All School Trip to Camp Mariah |
> | November | Parent/Student/Advisor Conferences |
> | | Advisory Bake Sale |
> | December | Festival of Lights, Pot-luck Supper |
> | January | Parents Association General Meeting |
> | February | Basketball Bonanza |
> | March | Parent/Student/Advisor Conferences |
> | April | Parent Association General Meeting |
> | May | Arts Night |
> | | Election of PA officers |
> | June | School Picnic |
> | | Prom |
> | | Graduation |
> | | Staff Appreciation Day |

the other kids and families in the school. This gave me an opportunity to interact with other parents. It is an informal gathering—everyone brings a dish. Parents have an opportunity to compare notes with each other—why they chose this school. It was also another opportunity for me to get a sense of the school community.... Again I was impressed by how well behaved the children were.

In addition, the picnic gets new parents connected with the parent organization, and they are all warmly invited to take an active role in the group. Two of the parents interviewed remembered how quickly they were included into the parent leadership group from their first meeting at this picnic. One parent talked about the difference at Crossroads: "I have had experience with schools where there was a strong sense of community. But I have also had experience with schools where the lack of community was evident. For instance, in one school my child and I were new to, the president of the parent association didn't introduce herself to me as we were both waiting for our children at dismissal. She continued talking to another parent and left me to myself." Such a situation would not occur at Crossroads because all new parents are invited into the fold. Another parent recalled that, even though she had not been active in her child's elementary school, she approached the PA president at Crossroads during the June picnic and said that she wanted to be involved in the school and asked how she might help. Not only was she invited to a summer planning meeting, she got so involved that, before the summer was over, she was designated to fill an open officer position of the PA. She said, "I felt like I was 'at home' with this group of parents. I felt that I had found myself a community. That's what I wanted." She became part of what was to become a particularly close-knit group of primarily older moms. Crossing racial lines, these moms found that they had a lot in common—they were seeking community and found it at Crossroads.

The Mission of the Parent Association: Building a Diverse Community

The 2000 U.S. Census shows that many schools across the country are becoming more and more diverse in terms of race, ethnicity, language, and class. Although suburbia and small towns used to be fairly homogenous, new immigrants are moving in. The major challenge facing schools in the coming years will be to help families of diverse backgrounds coalesce into school communities of difference. Crossroads already faces this challenge.

Given the diversity of its students, Crossroads is an example of what more schools will face in the coming years. Students' families come from Mexico, Haiti, Santo Domingo, Puerto Rico, and Eastern Europe; from different religious backgrounds—Jewish, Muslim, Buddhist, and a variety of Christian religions; from different racial backgrounds—African American, European American, Asian American; and from different classes—Upper West Side Jewish intellectuals and homeless families. Director Wiener says: "In this school we really have to work hard to get families to coalesce as a community. They [parents] relate to the school better than to each other. This is due to the fact that our families come from such diverse backgrounds in terms of race, ethnicity, language, and class and because they live throughout the city. Also many live in very stressful financial situations—leaving little time to attend to school issues."

Sidebar 8.4

Crossroads
School/Parent/Student Compact
Developed by the School Site Council
September 2001

Crossroads Mission Statement

Our mission is to provide an exciting, stimulating curriculum full of experiential learning that challenges all children to learn challenging content and to become accomplished students and lifelong learners. In addition, it is our goal to establish a democratic community of students, families, and staff that is diverse racially, ethnically, economically, and in ability and where all live and work together productively.

Staff
Provide quality teaching, advisory, guidance, evaluation, and leadership.
Hold students responsible for class work, homework, and good citizenship.
Offer respect for all people, ideas, creativity, nature, the environment, property, and the law.
Provide inspiration and humor, where possible and appropriate.
Remain open to communication from students, parents, and other staff.
Support growth and development of the school community and encourage development of student membership in the larger community of city, state, country, and the world.

Parents
Provide a homework environment that includes space to work, light, quiet, and basic equipment.
Require school attendance.
Attend parent/student/advisor conferences.
Be willing to help and to listen.
Provide expectations concerning academic effort, including homework, class work, and responsible school citizenship.
Provide guidance concerning personal responsibility.

Students
Participate in class.
Complete homework.
Have regular school attendance.
Attend parent/student/advisor conferences.
Learn good citizenship and personal responsibility.
Learn respect for all people, ideas, creativity, nature, the environment, property, and the law.

One especially complex issue has been the difficulty of recruiting more Latino parents to take an active role in the Parent Association. While Latino parents are "always there when you need them"—they attend school events, they attend parent conferences—they are not as active in the Parent Association. Cultural differences and language barriers may account for this. Attempts to help overcome the language barrier include having translators available at PA meetings, holding separate meetings in Spanish, and sending home all written notices in both English and Spanish. However, cultural expectations about the role of parents in relation to schools (described in a later section) are harder to bridge. What is encouraging, though, is that the parent leadership recognizes this as a dilemma and is actively working to make Crossroads' parent organizations more inviting to Latino parents.

Crossroads is working hard to be a community that is "savoring its differences and working hard to be nurturing to all."[21] The leaders of the Parent Association see their mission as one of community building. Parents are invited to be part of the community—to build relationships with each other—as well as with the staff through many formal and informal activities. Opening the school year with a pot-luck dinner meeting, the Parent Association is an active group that holds monthly meetings on topics related to the school program and parents' interests, such as reading, assessment, math, and discipline. These meetings are opportunities for parents to have open-ended conversations with the school director and other guests. They are, as one parent said, an opportunity for parents to "share their wisdom with each other" and to ask "really honest questions." The conversation about racism began during one of these meetings and then continued in the online discussion.

The Parent Association Steering Committee, an elected group of parents who are the backbone of the organization, meet monthly at each other's homes for a business meeting. Members are also connected via an online discussion group, proposed and set up by parent and computer enthusiast, Harriet Bograd. Subscribed to by more than 70 people including the school director, the steering committee (open to all), the school leadership team as well as anyone else who wishes to join, this discussion list provides the glue that holds together the organization, allowing for business and brainstorming to occur and keeping everyone connected to the conversation. While the larger parent community is not involved, the conversation among this group has forged a community that crosses racial, ethnic, and class lines. In addition to mundane business issues, such as when and where the next meeting will be, and to more substantive discussions like the ongoing conversation that was raised about race, discussion group members share personal stories and plan social gatherings with each other. For instance, as an observance for the Martin Luther King holiday, one couple arranged to attend another couple's church for their gospel service and responded in turn by inviting the other couple to attend a service at their synagogue—small steps toward understanding and appreciating each other's cultures.

One hallmark of a cohesive community is that community members take care of each other. Parent leader Harriet Bograd never missed an opportunity to take a grassroots approach to community problems. At one point, when she realized that some parents and students were not connected to the Internet because they didn't have computers, she organized a computer reclamation project. She recruited people to solicit donations of computers, which were reconditioned with the help of other parents and family members who knew how to do this, and offered computer workshops at her home. Then these computers were given to Crossroads families—taken home with the help of another parent who had a car. Another time, when some parents in the school had difficulty navigating the bureaucratic red tape of many of the institutions they had to deal with, Harriet proposed setting up an informal resource bank of parents who could help others deal more effectively with these bureaucracies. Although this project is not yet fully realized, it shows that Crossroads parents do reach out to help others through difficult times.

Parents Influence What Goes on in the School

Crossroads parents through their representatives on the School Leadership team—an equal number of parents to the number of staff members—have the ability to influence school policy, but their influence is even more apparent through informal channels. As feminist educational scholar Mary Henry suggests, a new paradigm of school organization encourages informal as well as formal opportunities for interaction.[22] Informal opportunities, occasions when people get to know each other as people, set the stage for creating ways to influence what happens in the organization. In fact, one of the reasons women have not risen in the ranks of management in larger numbers is that they are not part of the informal networks men have used for many years. In a similar way, lack of access to informal networks also disadvantages some parents in conversations about school practice.

Sidebar 8.5

Marjorie Moore, Parent
Crossroads Steering Committee Yahoo! Group
From October 1, 1999

Crossroads School Parent Statement

My kid is struggling with so many emotional things this year but, in spite of that, I saw the lights go on when she was given a seventh grade math problem the other night and she knew "how to think about it" in order to come up with an answer. Now even she admits that she can work on presentation stuff like neatness/orderliness, etc., but we hugged and kissed at this major math step. Will she be able to get a correct result when she applies her thinking to a specific problem? I bet not all of the time, but her chances have just jumped 90 percent. "How to think about things" math, politics, emotional ups and downs, getting along with others, how to relate to authority, team building, fulfilling her dreams, etc. . . .

Why does this mean so much to me? Because we are in a world right now without walls (Internet) and our children can literally learn and get information to do anything they desire regardless of race, sex, economic status, etc. I stress the "how" in how to think as "the" major skill of the next century. There will be so much information available to them (good and bad) that they will need critical thinking, intuition, and a host of other internal competencies like honesty, clarity, self-awareness. A major issue for me is "How will the world look when our children are in their twenties and thirties, and how are we preparing them for that world? I am convinced that our kids will actually see on the news extraterrestrial meetings of one kind or another . . . they will see the environment on earth deteriorate before their eyes and they will be thinking about how to survive without oxygen, contaminated water and food . . . and so on. I don't mean to be preachy but I think these are realistic projections based on how we have handled ourselves using outdated paradigms.

How is what we are doing now preparing them for a reasonable future?
What is learning fast enough in each person's life?
What is too slow, etc.?
How did teacher Jim Cole do that with her in the sixth grade? What was it about his teaching that made it possible for her to get it that night? What was the focus in the classroom? It wasn't homework because we didn't get a lot of that . . . it wasn't tests because they didn't get a lot of those. What exactly took place? What language do we use to describe it to others?

I could write the same or similar question about all of the other teachers.

To help parents and teachers develop a more specific language or style to talk about what it is we are doing . . . it is never easy to describe things like magic or the "ahas" but I think it is our destiny at Crossroads (smile).

Keeping relationships only at a formal level is problematic. When parents are kept at arm's length, when relationships between parents and educators are structured and controlled by educators, parents have little or no opportunity to have their voices heard. Even when they are included on decision-making councils, educators may not listen to what they say.

On the other hand, because Crossroads parents know and are known by the educational staff, they can influence what happens in the school. Of course, some staff members and many parents do not think that parents should have a vote about what the educational vision should be. The school has already developed a vision, so the goal now is different. As one parent said, "We have one that is strong. We need to help more parents understand it." However, she then added, "That doesn't mean that parents can't have input. Ann is very open to suggestions."

The discussion about Martin Luther King Day and Black History Month that opened this chapter illustrates just how influential parents can be. Parents believe that their ideas are listened to with respect—which is one reason they chose Crossroads School for their children.

BRIDGING CULTURAL TENSIONS IN A COMMUNITY OF DIFFERENCE

Just as problems occur even in the most stable families, all is not perfect at Crossroads. For example, tensions have emerged when parents and teachers of different cultures understand the school's practices—both its academic program and the advisory program—differently. Ann Wiener explains, "We are a work in progress—we continue to learn how to be more sensitive to each other's cultures." Because the staff and faculty have different cultural backgrounds from many of the children they teach and don't speak their first languages, they don't always understand how parents will respond to the decisions they make. And these cultural differences have led parents and faculty to see some issues in different ways.

The academic program has been a concern for parents of African American sons who, as one parent put it, "need to be more vigilant about their son's education to make sure they are prepared to cope in society." Parents worry whether the school's academic program adequately prepares them for succeeding in an imperfect society. They have asked many questions: Is there enough homework? How will my son do on standardized tests? Will my son get into a good high school? Is the school doing enough to prepare them?

Latino parents share some of the same concerns. For instance, one parent asked these questions about her child: "What was his level of capacity in school? His relationship with students and teachers? What was his progress? Really, how will most of this work apply to his future education? Would his work help him get into a good high school?"[23]

But some Latino parents also question the roles that teachers and parents play at Crossroads. Because many of the parents went to school in their home countries, they have very different images about what schools do and the relationships between teachers and parents. In many of their home countries the teacher was viewed as the child's "second parent," and parents trust teachers to do what's best for their children. To question a teacher may be viewed as a sign of disrespect.

Other parents, the school director observes, have concerns because "we don't teach in the ways they were used to." Parent Kathleen Malu, who conducted an evaluation of parent perspectives, concluded:

> Parents in this study never noted that they were asked about the concerns they had for their children's work in school. By asking such questions I believe that teachers may be able to help parents articulate their concerns, and the family conference might serve as an authentic vehicle for understanding children's learning and growth. . . . Lines of communication need to be reciprocal so that parents, too, can share their concerns and questions about their understanding of their children's work.[24]

Crossroads teachers, like teachers in other schools, have become experts about teaching and learning. They want parents to trust that decisions about curriculum and methods are grounded in their professional expertise—that they know what they are doing and that it is in the best interests of the students. On the other hand, parents, who have special knowledge about their own children and a vested interest in children's success, want to make sure that teachers are doing what they consider best for their children. Parents want to be included in the discussion about best practices.

In a variety of ways, teachers at Crossroads explain to parents what they are doing and why. Teachers have an opportunity to talk with parents about how they teach in both formal presentations to parent groups and during informal meetings such as those that occur at the many nightly events that both parents and teachers attend. However, since parents are not of one mind, teachers must find ways to take into account their many and diverse opinions. For instance, one teacher said: "I did a workshop on homework for the Parent Association because there are so many different perspectives among the parents. Some are upset because there is not enough homework; others because there is too much!" The workshop was clearly a challenge for the teacher but worthwhile in negotiating differences. As parents began to hear different perspectives from each other, they realized that there are many different ways to think about an issue. To build a community of difference, everyone has to learn not only to respect people who are from different backgrounds but also to respect those with whom they disagree. When a group has a common purpose—doing what is best for children—but there seems to be no common ground, then members must agree to disagree.

Another tension at Crossroads focused on the advisory program. Some parents perceive the teacher/advisor's interest in their child as an attempt to override the parent's prerogative. As adolescents begin to break away from their parents and try to assert their own independence, children often find it easier to talk with their teacher/advisors, many of whom are closer in age to them, than to their own parents. In addition, many teachers see their goal as helping these youngsters to take on responsibility for their own learning. These two dynamics work together to lead some parents—especially those from different cultures—to feel as if they are being excluded from decisions about their children. For instance, in one case related to choosing an appropriate high school, the advisor spoke to the student about her choices and filled out the student's application—leaving the parents to feel excluded.[25] The irony at Crossroads is that, although parents in other schools complain that teachers do not care enough, here some parents complain that teachers care too much.

One ongoing challenge for Crossroads teachers and parents is finding the right balance for parent responsibility, teacher/advisor responsibility, and student responsibility for the student's academic, social, and emotional wellbeing. Communication is critical to this process because everyone must make sure that all are kept informed about what each one is doing. As one teacher put it: "We are struggling to learn how to parent as a community."

Cultural differences can exacerbate the tensions between parents and staff, especially when students must choose high schools to attend when they leave Crossroads. For instance, one young outstanding Latina student talked to her advisor about this decision. The advisor, a young white man who was not fluent in Spanish, encouraged her to think of a boarding school, knowing that her academic prowess and leadership skills would win her a scholarship. Her parents would not hear of it and resented this teacher's counsel. Resolving the problem required the intervention of the school director and the parent liaison, a Spanish-speaking staff member who lives in the community. After the parent liaison served as a translator, the parents and the teacher finally were able to communicate and work out a compromise: The young woman would attend a private day school in the city.

WHAT DOES THE FUTURE HOLD FOR CROSSROADS?

Running any school is never easy or without problems. Running a school that is trying to break the mold and move toward a new way of thinking about schooling and relationships with families, as Crossroads is doing, makes the challenge even greater. Sometimes, as one parent put it, "things are a little discombobulated. Some ideas, administrative details, get lost in the hubbub of the day. But their hearts are in the right place."

Recently the school has made adjustments to try to help the director deal with administrative issues. For instance, the staff decided to reduce the teaching load of one teacher so he could devote himself to discipline issues. And, during the 2001–2002 school year as the school grew larger, a three-person administrative council was formed to give the school director additional administrative support.

These administrative changes, however, do not mean that relationships get overlooked. At Crossroads people come first. As the school continues to grow and mature, it is not likely that complacency will settle in because the school is a model for continued learning. However, as was the case with the development of the Code of Conduct, more processes and ways of doing things will become routine, and fewer loose ends will be lost. The learning culture helps in addressing administrative as well as academic matters.

Crossroads will also have to deal with external factors, such as New York City's budget crisis and competing interests for space in the 100-year-old school building, which Crossroads shares with two other schools. The budget crunch has already hit the school hard, requiring it to take more children than the staff believes it is adequately staffed to deal with and mandating the replacement of two staff members due to union rules related to layoffs and seniority. Director Ann Wiener and her staff are feisty and won't back down when challenged. They will put the needs of their students and families front and center and fight to get what they need. The parents, who have trusted their children to this school in which committed professionals care for their children, one by one, will provide backup support whenever it is required.

9

LINKING HOME, SCHOOL, AND COMMUNITY: A SAMPLING OF STRATEGIES AND PRACTICES

Earlier chapters presented a framework for thinking about a new paradigm for home, school, and community relationships as well as stories and profiles that illustrate the effect of these three spheres of influence on children and some ideas for getting started. This chapter offers a sampling of programs and practices from schools and communities across the nation.[1]

We invite readers to think of the chapter as a catalog of ideas for browsing because, although it is divided into four broad categories, in many instances the practices described in one section also illustrate another category—(1) Home, School, and Community Communication; (2) Families and Schools; (3) Parents in Schools; and (4) Home, School, and Community Connections. Several individuals contributed firsthand accounts of these connections in their own words. These are interspersed as sidebars throughout the chapter.

Excerpts from profiles written about South Side Elementary School in Johnson City, Tennessee; Kipps Elementary School in Blacksburg, Virginia; Discovery Elementary School in Buffalo, Minnesota; and Danebo Elementary School in Eugene, Oregon, introduce sections of the chapter. All four schools were recognized by the Boyer Center in 2001 for their exemplary work in building school communities that reach out to families and communities.[2] The nonprofit Boyer Center was established to build on the work of noted educator Ernest L. Boyer, Sr., who recognized that "[t]he circle of community extends outward to embrace parents, who are viewed as a child's first and most important teachers."[3]

This potpourri of ideas can serve as jumping-off places for schools and communities at different stages of building closer relationships with each other. Some require the commitment of significant resources; others, none at all—unless time is counted as a resource. Some practices may be good for one community but not for others. Each school or community must judge the benefit of a practice for its particular context.

All of the practices, however, reflect a commitment to open, authentic, two-way communication and suggest some ways that parents, educators, and community members can work collaboratively to promote the well-being of all children and educating them to be engaged citizens, parents, and workers.

HOME, SCHOOL, AND COMMUNITY COMMUNICATION

Too often, schools are islands separated from the families they serve and the communities in which their students live. Yet nothing is more important to the success of any relationship than communication. Despite their many competing demands and responsibilities, parents and teachers must find ways and times to communicate. The well-being of children requires that we try.

Home–School Communication

> *Communication is the key at Discovery. Each week the teachers send home newsletters that outline what the children will be studying and what needs the teachers have in order to make the curriculum work. Teachers and parents report that parents propose units they themselves might teach.*
>
> —The Boyer Center, *Boyer Best Practices 2001*

Both parents and teachers agree that good communication is an essential ingredient for a positive home–school relationship. Some say poor relationships stem from a lack of communication while others see its absence as a symptom of relationships that sour.

Effective home–school communication requires educators and parents to listen to each other. Educators need to find ways to hear what parents have to say and to act on their input. Parents often come away from interactions with

Sidebar 9.1

Kelly Arsenault and Debbie Deschambeault, Teachers
Lincoln Middle School, Portland, Maine

The More, the Merrier: A Successful and Effective Strategy for Building Bridges of Communication

Teaching 40 eighth graders and dealing with 80-plus parents and guardians often makes communication a daunting task. It takes teamwork and effort on the part of all parties involved to form the chains that link students, parents, and families together for successful learning. For the past two years we have worked as a teaching team at Lincoln Middle School in Portland, Maine. During our time teaching together we found a need to add new members to our team: parents/guardians. A number of strategies were implemented such as e-mail, Internet websites, notices, calendars, report cards, and phone calls. While all of these are effective, there was a noticeable need to bring everyone together.

Lincoln Middle School has been working on portfolio development as a means to connect the lines of communication between students, parents, and teachers as it relates to the educational process. These portfolios are designed to reflect the standards of learning outlined by the Maine Learning Results. Portfolios are divided into sections defined by the seven key Learner Expectations: collaborative worker, knowledgeable person, versatile thinker, self-directed learner, effective communicator, involved citizen, and quality producer. Throughout the school year, in designated class periods, work chosen by both student and teacher that reflects these standards is entered into the portfolio. Entries depict a progression towards mastery of the Learner Expectations.

Twice a year these portfolios are shared with community members, parents, and educators through student-led conferences. These meetings take the place of traditional parent/teacher conferences. Students are responsible for facilitating the discussion starting with simple introductions, to explaining where they are currently at in their progression of learning. Through this type of conference students not only work toward their goal of becoming effective communicators, but they also share success strategies and obstacles so that parents and educators may better understand their learning style. Because of their involvement, students place greater importance on the conference, thus resulting in higher parental attendance. This results in bringing all involved parties together in building a more cohesive team.

Feedback from the portfolio process and student-led conferences has been both positive and negative. Parents enjoyed hearing their child speak openly about his or her learning but preferred a variety of methods for students to share their output other than the traditional portfolio entry slip where written reflections were the only practice. Also, as students neared the second half of the year, their enthusiasm for the portfolio process declined. This was due to a number of factors including the repetitive nature of the entry forms, which a number of students had been using since the sixth grade. Varying the reflection process for both student and parents was becoming key for success.

(continues)

> **Sidebar 9.1** *(continued)*
>
> In order to accommodate all invested parties, we looked at educational goals, parental suggestions, and student interest and skills. It was unanimous that technology was to be a factor in the successful continuation of the portfolio process. Students spent the second half of the school year learning a variety of technological skills that could be incorporated into a PowerPoint presentation to be shared with a larger audience. The theme of their presentation was a reflection of their middle-school experience as it related to the Learner Expectations. Students used artifacts from all three grades, subject areas, and community and extracurricula activities as proof of their mastery.
>
> Parental involvement was important to the creation of the presentation as saved projects, certificates, trophies, photographs, and more were collected and brought into school to be digitally photographed, videotaped, or scanned. After the initial instruction of how to put the presentations together, each one took on a life of its own as students added sound, video clips, pictures, and color. Presentations were individualized, creative, entertaining, and reflective of the middle-school experience. We as teachers felt this truly was a celebration of what middle school is really all about, and students were eager to share their accomplishments.
>
> At the end of the school year, students invited their parents/guardians, community members, and previous educators to a pot-luck lunch that was followed by the sharing of their PowerPoint presentations. With great pride and enthusiasm, students presented their creations to the audience. The response of all involved was overwhelmingly positive. Parents were surprised and pleased to find what their children thought to be important aspects of their middle school years, as well as the creativity and talents their children possessed in communicating their achievements. Despite some initial student reservations, all involved enjoyed the day of sharing.
>
> Two teachers, 40 students, and 80 plus parents gathered to celebrate in a variety of successful modes of communication. Students shared their work and thoughts of their educational setting, parents had a dialogue with their children and teachers about learning, and all met in a positive, proactive atmosphere of camaraderie, cooperation, and communication.

schools saying that educators listened politely but nothing came of it. On the other hand, teachers hear from parents with many different points of view. Not all of their ideas can be accommodated in the school program. Creating a context in which people trust and respect each other and are open to learning from one another begins with communication. Research suggests that effective communication leads to increased parent involvement and student motivation, more positive parent evaluations of teachers, and higher levels of parent comfort with their children's schools.[4]

Parent leader Nicole Nichols-Solomon of Philadelphia suggests the metaphor of the flight crew of a commercial airliner for parent-teacher communication and collaboration: It takes more than one person to fly a plane. Communication is key: Everyone must be clear on the plan and what actions must be taken to avoid problems. Nichols-Solomon imagines a similar interdependent collaboration between parents and teachers.[5]

Many schools have begun to rethink the ways they communicate with parents. The sampling of ideas in this section suggests some ways to foster more open, two-way communication between educators and the parents of the children they serve.

Polling Parents in New York City
To find out what parents thought about their children's schools, the New York City School Board sent out a 35-question survey to parents. Most of the questions, printed in five different languages and answered on a five-point scale (ranging from "agree" to "disagree"), covered such topics as the competence of school principals, parent-teacher conference attendance, and how much homework children should have each night. School Chancellor Harold O. Levy explained why the survey was needed: "As the ultimate consumers of public education, the parents of our 1.1 million school children are in a position to evaluate how and what their children learn."[6]

Listening to Parents in Maryland
The education department of the state of Maryland created a statewide assessment plan for schools in 1991. The Maryland School Performance Assessment Program, as it was known, met with parental and teacher dissatisfaction throughout the state. Because there seemed to be serious misconceptions about the scoring, purpose, structure, and efficacy of the test, department officials responded.

Officials recognized that "our first challenge was to overcome teachers' feelings of disenfranchisement from the state department of education and their mistrust of the reform process." Thus, they enlisted teachers to help find out what parents' concerns were. In the process of the parent research, the department also learned more about and how to address teachers' concerns.

In response to both parent and teacher concerns, the department held over 500 meetings to explain the assessments to local districts, administrators, teachers, and parents. Officials also took a closer look at their language

Sidebar 9.2

Susan F. Bean
Teacher, Longfellow School
Brunswick, Maine

Connecting with My Parents

Every year we have a "Parent Meeting" very soon after the new school year begins. We give an overview of our program, talk a little about what we do daily and our routines, and ask for help. This past year, I took an idea from a colleague and posted a sign up sheet for parents to view during the talk and think about when and how they'd like to volunteer time, expertise and/or supplies.

Part of the talk has centered, lately, around Maine's Learning Results and how Brunswick has fine tuned those standards to meet Brunswick students' needs in a document called "The Brunswick Frameworks." I don't go into much specific detail except to point out that science is not my major and I am not really a scientist. This year I was brave enough to ask if there were any parents willing to do "hands-on science" with my kids. Amazingly, four people signed up!

The Science Experiments
We talked briefly after the meeting and agreed that I would send home the specific standards and performance indicators so that experiments could be developed around them. I had no real expectations except that parents would do the best they could and I would need to fit them into my weekly schedule.

We decided that Friday mornings between 10:00 and 10:40 would be the time set aside. It seemed most appropriate because my class would be out of the room for the hour prior to that meeting and partner time, and the experiments could be set up without all those "inquiring minds" asking questions! (I can never say, "Please do such and such while I set this up" and feel confident that the kids will ignore me until I'm ready!)

These experiments were successful beyond my wildest dreams! Not only did I not have to procure the supplies or set them up, I had real science people who knew how to develop experiments, research them on the Internet or wherever, and conduct them! I was more than happy to record on the chart or connect to what we'd been studying and, of course, provide students with the needed reminders to focus, etc.

We had experiments on energy—all kinds! We made a pseudo-phonograph out of paper and a toothpick; we used heat lamps and moved things with dry ice. We made layers of the ocean with gummy worms and Karo syrup and other ingredients to see what happened if there was too much plankton on the surface. We dissected mollusks. We tested for fats in foods; we tested for acids and bases, too. We had pet chickens visit. We froze things with liquid nitrogen and made ice cream which we ate!

For each experiment, we tried to maintain the steps of the scientific method: Question, Observations, Hypothesis, Experiment, Results and Conclu-

(continues)

Sidebar 9.2 *(continued)*

---

sions. We became so familiar and I so comfortable, eventually, that we were able to design and conduct our own experiments/questions about the forest. I hope I'll be as lucky this year!

The Weekly Newsletters
I have sent weekly newsletters every week for the past twelve years. I have evolved in my philosophy from including just basic information and upcoming events to using that forum for a real communication venue. I have, as anyone who writes a weekly column would, developed an arena for saying what needs to be said (in my opinion) and used my "voice" to connect with parents and colleagues.

At first, while I am getting to know my group and parents, I include that basic information and "need to know" things, reminders, and upcoming events. The hardest part is to get the parents to EXPECT those letters EVERY WEDNESDAY so that when they don't come home, they ask about them. Many parents have written in the assignment notebook on Wednesday night "I didn't get my newsletter! Where is it??" Then we fish it out of the "black hole" of a student's desk and hope the student won't forget again.

As the class develops a personality and the parents become known to me, I use the weekly newsletter as a chance to give feedback on what our class is learning or problems we are solving. (One year there were way too many love notes so we decided together how to eliminate them!) I tell them about projects we get involved in and celebrations at our school or in our grade level. I even ask for their help.

Many times I use the newsletter to "educate" parents about the Learning Results or the Writing Process or the Scientific Method. There might be practical suggestions about how to get fourth graders to do homework or their Home Reading. I give them descriptive information about how we get involved in learning these things and feedback about how they've succeeded (and how they fail once in a while!) Once in a while, when I'm looking for a "column" on Tuesday night, I use that space to clarify my own thinking and work things out "with an audience" to "keep me honest." I may commit to things in the newsletter, which makes it more likely that I will follow up on them. Often I include student reactions to questions or feedback on special events, but mostly it's a place for my views, my ideas and my writing.

My main purpose is to get parents to understand me and what I am trying to do with their children: to help them be clear and effective communicators, self-directed lifelong learners, creative and practical problem solvers, responsible and involved citizens, collaborative and quality workers and integrative and informed thinkers. I want parents to feel they know me well enough to be able to call or come in with a problem before it gets huge and takes longer to solve. Working together with parents in a partnership is the goal, and these letters seem to help.

and communication strategies, realizing that they, too, were barriers. For instance, when teachers learned that parents disliked the format of a pamphlet created to explain the exam, changes were made that appeased both state officials and parent organizations.[7]

Dial-A-Teacher Voice Mail in Aloha, Oregon
Teresa Jo Clemens-Brower, a teacher at the Errol Hassell Elementary School in Aloha, Oregon, created a Dial-A-Teacher Voice Mail system that allows parents and students to listen to the daily message and leave messages of their own. Clemens-Brower's message, about one minute in length, tells callers what is happening in class on that day. The message also highlights skills the students have been working on and provides information about homework and future plans.[8]

Sidebar 9.3

Julie Ann Ludden, English teacher
Windham (Maine) High School

E-mail: A Powerful Way to Communicate

Using e-mail has had such a positive impact on my communication with parents and guardians. Sometimes it only takes one message to "awaken" a parent to discuss school performance with the child. As a freshman English teacher, I often find students who enter high school and don't meet the expectations/requirements to pass. In the middle of the first quarter, I had two young ladies failing. They were not handing in homework, not reading independently, not studying. After e-mailing the girls' mothers, what a change! The young lady who "hated to read" transformed into a reader because her mother loved reading, and they started shopping for books together. Without the communication, I strongly believe she would have failed for the year.

Another young lady was also able to turn things around after I "talked" with her mom. This mother started studying with her daughter. Because I took the time to contact this parent, I connected with my student by showing her I cared. Now she's a sophomore and stopping by my room frequently to visit. I am so glad I took the time to type a message. I could also share in her excitement at the end of the year when she earned an A.

Josh was not only failing, but he was also a behavior problem. He lived with his aunt and uncle. The aunt and I e-mailed on a *daily* basis. Again, a bond was formed because this student knew I was concerned and cared. If he was ever disruptive, I'd smile and say, "Josh, I have two words for you: electronic mail!"

> **Sidebar 9.4**
>
> *Larry DeBlois*
> *High school English teacher (retired)*
> *Augusta, Maine*
>
> ## Mini-Schedule
>
> In September, about a month into the new school year, came time for the mini-schedule at my school, a combined 7 to 12 rural school. On a Tuesday night from 6:30 to 9 P.M., parents came to school, picked up the class schedule of their child, and spent 10 minutes in the classroom with each teacher who taught their child each particular subject.
>
> The evening began with coffee and goodies. The principal was there to meet and greet the parents. Members of the student government handed out the schedules and classroom maps (we had seven portable classrooms placed around the main building) and served as tour guides for any of the terminally lost.
>
> At 7 P.M. the parents met with their child's advisor (see sidebar 6.8) for 10 minutes. I gave the parents of my advisees an update on the plans for end of the year, a full fun day off campus, because we needed their help for transportation. I answered any questions they had or wrote down the ones I couldn't answer then. I told them I'd call them back when I found out the answers and invited them to contact me if they had further questions.
>
> Then the parents were off to the same classes their children took. The parents had about five minutes passing time. They got to class, received a syllabus from the teacher, and heard an overview of the class and its expectations. Time was left to answer a few questions. Then the bell rang and the parents had to hurry to the next period.
>
> At the end of the evening, the parents had an overview from the instructors of all the courses their children were taking, along with a syllabus of each of the courses. The parents got a chance to introduce themselves to the teacher and to see her or him in action at least for 10 minutes.
>
> We found this mini-schedule to be superb for our school. It was heavily attended, and the parents left feeling upbeat about the courses and about the enthusiasm of the teachers. Since teachers are all hams, most enjoyed the interchange with the parents. I always noticed some parents taking furious notes. Those whose children had been in my courses before just nodded their heads. It was high theater and great fun and worthwhile for all.

Parent Conferences Outside of School in Vinita, Oklahoma

Dianna Just, a high-school English teacher who works part time at Wal-Mart, uses this job to connect with her students and their parents. Although the conversations in the store are brief, she lets parents know about a student's progress and reminds students about quizzes or essays or which supplies they need for class. Just notes that she averages two "conferences" a night at Wal-Mart. Parents and students alike think she is more accessible at Wal-Mart than at Vinita High School.[9]

PTA Reaches Out to All Parents in Mt. Vernon, Virginia
Due to the determined efforts of its president, the Mt. Vernon High School PTA grew from a dismal turnout of 8 to 20 members per meeting to an outstanding 100 percent of teachers and a dramatic increase in the participation of parents and students. Recognizing that parents want to be actively involved in areas that most directly affect their children, the president reorganized the PTA, created a new mission statement, established a parent council in each grade, and developed community resource teams to distribute monthly newsletters and to gather support from the entire community.[10]

Parent Conferences Held in Neighborhoods of Cleveland, Ohio
Instead of assuming that parents who don't show up at school are not interested, Buhrer Elementary School in Cleveland, Ohio, makes it easy for parents to get involved in their children's education. Teachers hold parent conferences off campus in places closer to students' homes. The school also holds "Block Parent Meetings" for those families who cannot attend school events because they live on the outskirts of the community and lack transportation. Block meetings, which take place every few months in a parent's home or nearby library, address parents' concerns and offer an opportunity to discuss school-related information.[11]

Communicating with the Larger Community

As we've noted before, many people in a community have few links to schools—these include, single people, married couples without children, and older parents whose children are no longer in school. Others who work in the local businesses, nonprofit organizations, cultural organizations, and places of worship do not live in the community. Yet schools affect them. Their attitudes toward the schools affect the schools—especially when school budgets are debated and decided. Smart schools pay attention to communication with their communities as well as with parents.

KIVA: Engaging the Public in the Education Reform Dialogue
Connie Spinner of the Public Education Fund Network borrowed KIVA, a process for problem-solving, from the Southwest Native American Pueblo

tribal tradition to bring community members together for meaningful dialogues about school reform. She says: "We're guilty of not really listening to each other. We're already thinking about how we're going to respond to another person before they've finished talking. We don't take the time to process what they've said. KIVA sets people up to do just that."

The KIVA format is not complicated. Sixteen community leaders, representing students, teachers, parents, administrators, school board members, and community/business leaders, are invited to participate as speakers. Other community members are invited to be listeners. Each speaker receives three questions before the meeting and prepares responses. One by one, each speaker speaks in turn. Listeners then report one thing they have heard and one thing that they didn't hear from each speaker.

This structured approach to speaking and listening encourages active listening and critical reflection from everyone. Spinner points out, "This is not a soapbox. This is your chance to represent your group on this question in a thoughtful way."[12]

Common Ground Thinking in Modesto, California

School communities everywhere include parents, teachers, and community members with very different perspectives about what makes good schools. Finding ways to talk about our different perspectives without rancor can be very hard, as the Modesto School District discovered, when it faced a community protest for including "sexual orientation" in their new tolerance policy. The school district sought ways to disarm this conflict and to bring people together to share their different points of view in a safe and respectful way.

"Common ground thinking" formed the basis for these dialogues. According to Wayne Jacobsen, who was brought in to help the district do this, "Common ground thinking provides a legal and historical map through the minefields of the so-called culture wars." The advice he gave the school district was to face issues head on; invite all stakeholders to be part of the committee charged to address the issue; get training for committee members in a process that can lead to consensus; formulate and approve a policy agreed upon by all committee members; and train the staff and the community to understand the process. Community members who served on the committee played key roles in helping their constituencies to understand what they had developed.

The value of an approach like common ground thinking, is that it "removes educators from arbitrating social conflicts; . . . helps people appreciate and apply religious neutrality; ... switches the dialogue from what I want for my child to what is fair for all children, including mine; ... eliminates confusion, suspicion, and anger generated by advocacy groups; and ... recognizes the priority of the family in faith and values."[13]

FAMILIES AND SCHOOLS

Kipps Elementary School's attention to the diversity of students' socioeconomic status and the varying educational levels of the families means they consciously incorporate the needs of all the students in making plans and decisions. A book by Ruby Payne, A Framework for Understanding Poverty, *has been an important resource.*

—The Boyer Center, *Boyer Best Practices 2001*

At Danebo School opportunities for parents to meet and talk with each other include kindergarten group evening meetings once each month and some teachers hold classroom "family" meetings.

—The Boyer Center, *Boyer Best Practices 2001*

When parents and educators work together for the common purpose of children's well-being, they have a much better chance of protecting and educating all children well. But schools can support what parents do at home only when they have a deep understanding of and respect for what parents believe and value. Otherwise, children suffer. For example, Philadelphia parent facilitator Nicole Nichols-Solomon notes, "Low-income African American and Latino children are often perceived as limited by their socioeconomic status. Influenced by newspapers, television, and movies, many well-meaning teachers imagine their students fighting their way to school through a community of zombie crackheads and returning to homes that lack caring parents or other adults."[14]

The reality for these children and others, however, may be very different from teachers' perceptions. Even in the most impoverished communities, parents can and do provide many positive learning experiences for their children, which educators could build on if only they knew about them. And most parents want to help their children succeed in school, if only they knew how. Some schools have found some good ways for parents and teachers to learn about each other.

Cultural Interchange between and among Parents and Teachers

Everything we know about good teaching suggests that teachers must fashion an educational program based on the prior knowledge and experiences of children. Teachers, then, need to know more about who the children are outside of school. What have they learned from their families, from their extended families, from their neighbors, from their culture, from their religion?

Parents, of course, have this knowledge and can share it with teachers—if asked. Luis Moll and his colleagues at the University of Arizona in Tucson did just that. They involved their graduate students in a study of what Latino families in South Tucson know and teach their children. Visiting homes and neighborhoods, students found that these families had a tremendous wealth—"funds of knowledge"—to share with their children, all previously invisible to their teachers.[15] When working with families of different cultures, researcher and writer Susan Swap suggests that schools "meet with families in their communities . . . ; plan informal opportunities for contact and building trust . . . ; identify a liaison person . . . ; ask parents about their needs, interests, and priorities . . . ; base programs on parent and educator priorities and needs . . . ; develop clear guidelines about how parents can support their children's learning . . . ; offer options for parent education and support . . . ; develop theme-based curriculum units that draw on cultures that you are learning more about . . . ; collaborate with colleagues . . . ; and explore your own cultural values and assumptions."[16]

As the following examples show, teachers across the country have found ways to discover the "funds of knowledge" families have provided to their students.

Parent Stories about Home Literacy Practices
To aid teachers in gaining information about the home literacy environment of their students, Dr. Patricia A. Edwards and Heather M. Pleasant of Michigan State University's College of Education studied how parent stories might help. Parents were asked to tell stories to researchers about what they do at home to encourage their children to read and write. The study, focused on

> **Sidebar 9.5**
>
> *Kathy Gregg, Brendan Dundas, Nancy Fowler, and Sue Manning*
> *Pacific Seventh-Grade Team*
> *Fowler Middle School, Tigard, Oregon*
>
> ## Bridges
>
> "I'm taking away all privileges until you finish your homework!"
>
> "Mom, this is sooooooo boring. School sucks!"
>
> "I've just gotten off the phone with Mrs. Jones, and, once again, you have fallen behind. What do I need to do to get you to understand? This is important!" The only sound was the silence of frustration that settled between us.
>
> As a parent, I found myself in a place I'd been many times before. The familiar confrontation, the door slam, and the ensuing silence. I agonized late into the night. My thoughts centered around how I could get past this daily struggle. We were working at cross purposes. Patrick, as a seventh grader, was looking for the new independence of early adolescence. While I wanted to extend this to him, I also knew my role as a parent necessitated my consistent watchfulness and guidance. How could I reconcile out individual needs? What needed to happen with Patrick, me, and others in Patrick's life to help with this? I couldn't be the only parent faced with this situation because, as a teacher, I knew many of my students' families struggled with this same dilemma.
>
> Current research shows that children at the middle-school level want adults to be involved in their education, and, in fact, are more successful because of it. Knowing this, I decided to go to my teaching team and develop a set of ideas to bridge this all-too-familiar gap between our students and their parents. Due to flexible scheduling in our school, the team meets several times a week. In addition to regular schoolwide issues and concerns, we now include conversations that specifically address parent-student involvement.
>
> As a team, we realized that we had already made ourselves accessible in a variety of ways. E-mail, letters, phone calls, and meetings were currently part of our repertoire. Now we wanted to develop even more meaningful connections between home and school. Here are some strategies we developed and have successfully used.
>
> - *Parent/Student Questionnaire*. Parents are asked to tell us about their children, and students, in turn, are asked to tell us about their parents, for example. What is one of your all-time favorite books? What do you like to do for fun? What is great about you? (Bragging is encouraged.)
> - *Student Letters*. Informative letters from the teacher about what's going on in the classroom are great, but why not ask the source? We have students write this letter. In addition to simply telling parents about upcoming projects or due dates, we have students write about what's been happening with their learning, specific problems they are having, or an activity that was interesting or fun.
>
> *(continues)*

> **Sidebar 9.5** *(continued)*
>
> - *Connecting Reading and Writing.* Besides having students write about their reading and writing in school journals or reading logs, we also ask parents to respond to what their children have written.
> - *Home Reading and/or Writing Survey.* What kind of reading/writing happens at home? We ask students to survey all the types of reading and writing activities that they see going on in their homes. This activity helps students come to the realization that reading/writing is everywhere, not just in the classroom.
> - *Recipe Project.* When they study fractions, we ask students to find a favorite family recipe. To involve the parents, students ask them to make the dish together as a family.
> - *Student-led Conferences.* Parents are invited to view their children's work and listen to them explain specific successes in each content area. Because students rather than the teachers are the facilitators, parents have the opportunity to interact with their children in a personal setting.
> - *Project Night and Field Trips.* These are great ways to connect school and home, but the key is having students—not the teachers—extend the invitation to attend and explain the event.
>
> As a teacher, I know how easy it is to get caught up in lesson plans, grading papers, and discussions about school issues. Parents are often the forgotten element in the classroom experience. Our team is committed to continue building a better bridge, stone by stone. Already we have witnessed signs of success, such as when a student expressed excitement with the realization that a parent cares and when a parent calls to praise our consistent communication. We have chosen this path, and our journey is ongoing.

at-risk students, was geared toward providing teachers with a clearer picture of each individual child's needs. Parents' personal stories also help reduce the gap between parents and schools and can be a positive way for parents to participate in their children's learning. Schools, however, must consider the ways in which this personal information will be used because everyone's right to privacy must be upheld.[17]

Home Visits in Sacramento City, California

In order to strengthen ties between teachers and parents, the Sacramento City Unified School District implemented a program that involved teachers in visiting the homes of their students. Principals of schools participating in the program observe "better student behavior, improved homework quality, and closer, more open relationships with the communities they serve." Parents and teachers both speak positively about the program. One parent says, "I sometimes feel excited or nervous at school. It's better to speak at home because the conferences go too fast." And a teacher notes: "I appreciate and understand my students so much more."

Teachers have seen an increase in the test scores of their students since the home-visit program started. During these home visits, they show parents strategies they can use to help their children perform better academically. The home visits help teachers build stronger relationships with the parents. In order to ease anxieties and encourage teachers to participate, they receive training on what kinds of information to provide parents and ways to talk effectively with them.[18]

Parents and Their Children Coauthor Books in San Francisco
Parents coauthoring books with their children can open the lines of communication between parents and schools. According to researcher Sudia Paloma McCaleb, "If teachers welcome and validate parents by listening to their concerns and finding positive things to communicate to them about their children's ongoing progress, then the parents will most likely be open to a partnership."

McCaleb invited parents of first graders in one school in San Francisco to engage in a dialogue about their childhood experiences with education and their current roles in the educational process of their children. She helped parents to coauthor books about their stories with their children. Among the themes chosen were childhood friendships, families building together, and families as problem solvers. McCaleb observes that the process of coauthored books provides "literate occasions for parents and students to work together."

The project fostered communication within the family and community and led also to an increase in children's respect for their parents (and vice versa). The teachers got to know more about families and the community and what people thought about educational matters. The project also revealed a disconnect between parents and teachers about parent involvement. One parent commented, "I believe that the most important education of a child depends not only on the school. True education begins in the home and comes not only from the family but from the community in general." Another parent said, "When I have expressed my opinions, they have helped me but, in reality, there is no partnership. I follow what the school tells me to do, what the school asks of me. It's as if I were also a student in the school."[19]

The Family Photography Project in New York City
Photography was the vehicle of communication in a project created by Jane Spielman, a researcher and teacher at the City College School of Education in New York City. She invited nine Latino parents to take pictures of "moments of learning" in their families, explaining "I wanted the families themselves to define the ways in which they taught their children at home and in their community."

Parents took over 1,000 pictures from which they created photo essays. Parents then helped to combine these individual photo essays into a set of 93 photos, thematically organized into categories, such as love, friendship, and family; growing up to become brave and independent; responsibility and routine; culture, ritual, religion, and play; literacy at home; technology; and the science of learning in the community. This project helped parents to recognize just how much they taught their children. As one parent said, "I saw my values defined through the pictures." Another said, "I talked to my children with pictures about the values that we teach, both old values that we have already been teaching and new ones I want to teach."

Parents presented the finished photo essay to Spielman's class of practicing teachers. The teachers learned directly from parents what they [the parents] taught their children. Spielman describes the event: "The large classroom turned into a museum for the night as students [teachers in her class] wandered through the captioned photos taking notes. Studying the photos and meeting the families helped the teachers consolidate their inquiry projects and confirmed that families were the 'primary teachers' of their children." A parent summed the project up this way: "We educated the teachers so they can educate their students. . . . I never realized I could do such a thing."[20]

Mentor Parents Help Teachers in Stockton, California
"Mentor parents," trained at the district's parent resource center, spent 5,000 hours in the schools helping school staff improve family–school communication and parent involvement in their children's learning. Among other activities, mentor parents conducted four workshops on obstacles to parental involvement in schools. The workshops focused on such topics as parents' own negative experiences with school and teacher bias, which may result from cultural or language differences among teachers and parents.[21]

Helping Parents Help Their Children at Home

One important research findings is that what parents do at home to support the school's efforts is a major influence on how well children do in school. Most parents want to help their children. Some don't know how. Many schools are now reaching out to parents who need assistance to help their children at home.

Family Math

FAMILY MATH is a family-based program to encourage girls and minority students to enter careers that use mathematics. The six-week program, developed by EQUALS at the Lawrence Hall of Science at the University of California, Berkeley campus, brings parents and children together in groups of approximately 25. Parents and children learn math-related games the family can play at home. Parents discover many ways they can help their children with their education. By demonstrating new approaches to teaching mathematics, the program works to change old habits and attitudes toward mathematics.[22]

School Wide Interactive Homework Project in Enfield, Connecticut

"Family Connection" at the Eli Whitney School in Enfield, Connecticut, is an interactive homework project. At a staff meeting the teachers create a voluntary homework assignment. Topics have included "About My Family" and "Favorite Family Fun." Students create pictures or complete writing assignments with their parents to share in the classrooms and display in the hallways. These interactive projects are "low cost" and "nonthreatening" with "strong interest from parents." The project, planned to improve students' writing skills and parent-child interaction, also helps educators get to know the families of their students, improves school-to-home communication, and increases the number of volunteers. The project celebrates and shows the school's appreciation for its families.[23]

The Chicago Public School's Parent and Community Training Academy

The Parent and Community Training Academy (PCTA) helps interested parents improve the academic achievement and social development of their children. Through a curriculum that integrates writing throughout all subjects areas and grade levels, parents develop a better understanding of what their children learn and how they can help their children become more skilled readers.

The PCTA provides training for parents in many other areas, as well, including health and nutrition information, safety and violence prevention, parenting and life-skills information, and character education training. The character education training stresses traits such as integrity, honesty, good sportsmanship, healthy minds, and positive attitudes that the parents can reinforce with their children at home. The PCTA also refers parents to adult education and General Educational Diploma (GED) courses.[24]

Red Bag Helps Migrant Families
Because migrant families by definition do not stay in one place very long, children often transfer from one school to another. The new schools need academic and health information that may be difficult for parents to maintain. In one area with many migrant families, educators came up with a simple plan to help parents deal with these bureaucratic requirements. In addition to making sure that parents knew what documents would be needed each time children enrolled in a different school, they gave parents a red bag to transport all the information to the new school. The red bag "serves as a tool for appropriate student placement, as well as a means of empowering migrant parents to ensure the educational progress of their children."[25]

The Parent Connection in Oregon, Mt. Morris, and Chana, Illinois
A demonstration site of the Illinois Family Education Center, the Parent Connection works primarily through the five district schools (high school, junior high, and three elementary schools) but also assists the area's 10 preschools and collaborates with community agencies, churches, and youth organizations. The program provides parents with education for leadership, parenting, and child development. It also facilitated the development of School Community Councils based on the Alliance for Achievement (another division of the Illinois Family Education Center) blueprint for building school communities.

Sally Weber, coordinator of the program who enjoys a close relationship with the district's administration and faculty, says, "I am called upon regularly to represent the parent's interests in situations in our schools.... My staff and I help with parent-teacher conferences. I work with other school personnel to visit the homes of incoming freshmen. Teachers and counselors consult with me on family issues and refer parents to me who need assistance and support. All of this family support is a part of the service of the Parent Connection that is in addition to the training we provide."[26] One of the programs offered through the Family Connection and in use in the Oregon and Mt. Morris, Illinois, schools is "Families and Schools Together" (F.A.S.T.)

F.A.S.T., a nationally replicated parent involvement program, engages families in enjoyable, research-based activities. Developed by Lynn McDonald and Family Service of Madison, Wisconsin, its goal is to strengthen families and empower parents.[27] The program is implemented by a set of school and community partners: a mental health partner, an arts and crafts partner, a coordinating partner, a parent partner, a substance abuse partner, and a school partner. The families meet once a week for 6 to 8 weeks, participating in arts and crafts, outdoor activities, and a dinner. Some benefits of the program are improvements in children's behavior and family functioning and family-school involvement, a decrease in the parents' feelings of stress and isolation, and a willingness on the part of the parents to provide the schools with more information for school records.[28]

Teaching Parents English as a Second Language in Logan Square, Illinois
The fact that parents who don't speak English well have difficulty communicating with their children's schools creates an educational barrier. The Illinois Family Education Center (IFEC) offers English as a Second Language (ESL) courses in its schools. The Mozart and Goethe Schools in Logan Square offer ESL for Effective Parenting classes. The instruction, which includes the use of role-playing, photos, flashcards, and real objects, is based on the needs and desires of the adult students in the classes.[29]

Helping Parents Help Their Children Prepare for College in Dallas, Texas
Roosevelt High School in Dallas, Texas, makes a special effort to help parents understand what their children need to succeed in school and in life. The school invited parents to an evening class to review the state assessment instrument and to discuss the skills their children are expected to demonstrate on the test. Plans are under way for staff to train parents to help students develop the skills necessary to apply for college, such as completing financial aid forms, obtaining references, and preparing for required standardized tests.[30]

PARENTS IN SCHOOLS

Discovery School is clearly an exemplary model of parent and family involvement. In fact, there is such a symbiosis between

> *teachers and parents it is difficult to tell them apart—they function as instructional teams. Parents working in the instructional program are visible throughout the school; younger siblings are welcome to come and participate along with their parents.*
>
> —The Boyer Center, *Boyer Best Practices 2001*

In the past, parent participation in schools has been limited to attending performances, PTA bake sales, going on field trips, and running off copies for teachers. Many teachers now recognize the value of involving parents in more substantial ways. Parents tutor children, help with small-group instruction and special projects, make presentations in classrooms, sit on advisory committees and governing boards, attend classes, and, in some cases, even participate with teachers in professional development activities. Although parent participation tends to decrease as their children move to middle school and high school, some schools have successfully involved parents in the upper grades. Creating a welcoming environment is a good first step.

Creating an Inviting Environment

> *Danebo Elementary School takes pride in its welcoming environment. One relatively new mother to the school speaks of the caring and helpful greetings she received on the first day of school.*
>
> —The Boyer Center, *Boyer Best Practices 2001*

> *People at South Side feel that the school's staff demonstrates a sincere commitment to carry on the school's tradition as the community's valued focal point. They feel welcome in the school; everyone from the secretary to the custodian knows them and their children.*
>
> —The Boyer Center, *Boyer Best Practices 2001*

When visitors—or parents and students new to the school—enter unfamiliar schools, they can tell right away whether they are welcome. In one school, the icy cold atmosphere of a hallway is devoid of anything but the legal notice "All visitors must report to the office." In another, the hallway is filled with children's artwork and a sign that says "Please help us provide a safe environment for our children and report to the office for a visitor's pass." A teacher in one school passes the newcomers in the hallway without a nod whereas another asks with a smile, "How might I help you?" One school secretary pays no attention to the newcomers when they enter the office, but another offers a warm welcome. In one school, the visitor sees a teacher shout at a student and in another, a teacher chatting with a student as they walk together.

Parents and others come to schools with different past experiences. Some have had bad experiences as students; others dropped out, perhaps without learning to read; and others learned first to speak a language other than English. What do schools do to help all visitors feel welcome, to sense that people in the school care about children and want adults as contributing members of the school community?

Many educators are doing a great deal that goes well beyond creating a positive first impression to make their schools warm and welcoming places.

School Development Program: Creating Caring Communities in School
The School Development Program, a national program adopted by more than 650 schools, models how educators can work with families to create caring communities in schools. The program's founder, James Comer, first went into schools to figure out what was wrong with them. He says, "We realized that the children were bright and able but that the climate wasn't right. We also realized that the teachers wanted to succeed, but they were stuck with a mechanical model of teaching and did not understand what else was necessary."

Comer, along with others, developed a structure that allowed parents, educators, and other specialists to develop a comprehensive school plan—one that would create a positive learning environment for children and adults. To create a welcoming environment they developed activities such as a "Welcome Back to School" pot luck supper. Comer says, "Every school must become a place that supports the development of children ... because social and emotional growth are so important to academic learning." He also notes that schools can help address the breakdown in the sense of community people have. "The solution is to restore a sense of community. And that is what our program has always been about: restoring community, and doing it within the school."[31]

First-Day Celebrations in Schools across the Nation
An increasing number of schools are starting the first day of classes with "first-day holidays." In many schools, the first day has turned into an occa-

sion that involves both parents and students in various activities to get to know the school, the teachers, and other students and parents. Employers are encouraged to allow their employees to take time off to participate. Terry Ehrich, from Bennington, Vermont, president of the Vermont-based First Day Foundation whose town started the traditions of first-day celebrations, hopes to get schools across the country to join in the festivities. When parents feel they are partners in their child's education from the beginning of the school year, they are more likely to volunteer and continue their presence in the school beyond the first-day activities.

First-day activities are not just social events; they also build partnerships among the schools, the parents, and the communities. "Schools are encouraged to organize their events around a particular goal, such as expanding after-school programs or modernizing their buildings." Activities include parades, picnics, traditional school lunches with parents, and question-and-answer sessions for parents. To make all parents feel welcome, one school also created a Spanish-speaking parent club for parents who may not yet feel comfortable enough to participate in the regular parent club.[32]

Parents Volunteer in School

> *Classroom visits find parents and community persons working alongside teachers in classrooms, hallways, learning centers, and wherever space can be found. These volunteers display a sense of autonomy as they work with children in small groups. They know the children by name and how to work with them without direct supervision. Full-time mothers, a local attorney, working moms and dads who take time from their jobs volunteer as reading tutors. Grandparents, retirees and community business persons—all work in various capacities in the school.*
>
> —The Boyer Center, *Boyer Best Practices 2001*

Some teachers admit they could not run their classrooms without the help of parents. They depend on parents to help with special projects, manage the daily routines, and tutor children. As with every good idea, however, there are some downsides to parents as volunteers that must be managed. Some teachers notice that when parents are in the classroom, some children become more immature and cling to their moms. Some parents admit that one

Sidebar 9.6

Pat Monohan, First-Grade Teacher
St. Agatha Acadmy, Winchester, Kentucky

"Mom and Pop Reading Day"

When I announce it's " Mom-Pop" Day, cheers ring out in the classroom. " Mom-Pop" Day gives parents, other family members, and friends of the family who come to my classroom a chance to help children with reading. Here's how it works. Each adult draws an index card with a child's name on it and listens to that child read for about 10 minutes. The adult records the date, the book the child reads, and comments for feedback. Then, each draws another card and so on with different children for the rest of the 40-minute reading time. Parents usually spend the last 10 minutes with their own children. We have these special days once a week because the students really enjoy spending time with adults.

Julie Macosko Kerber, Parent
St. Agatha Academy, Winchester, Kentucky

"Mom and Pop Reading Day" from a Parent's Perspective

What a joy for me to hear my child say that reading is her favorite activity at school! And what a gift " Mom and Pop Reading Day" is for adults and students every Wednesday morning at St. Agatha! We adults come to the classroom not to read to the students but to *listen* to the students read. And during this open reading time the children may sit wherever they want. On the floor with pillows is a favorite spot for many.

So, often, as I've sat listening to the children read, I've recalled my own first-grade year. I can still remember, even after more than 30 years, being excited about reading. It is such a privilege to be part of that process in my own daughter's life and also in the lives of her schoolmates. I like the idea that my children have other adults in their lives, caring about them, listening to them, and just being there. "Mom and Pop Reading" allows for that. I appreciate seeing how different the children are, how differently they interact, and how varied their reading levels are. An added treat—and helpful to those of us who can't always find child care—is that younger siblings are invited to come with their parents. Having the little ones there fosters the value of reading and makes for special sharing time.

of the reasons that they volunteer is to "keep an eye" on what happens to their children at school. In both cases, open and honest communication between teacher and parent can allay concerns.

Operation Bookworm in South Brunswick, New Jersey
Reading teachers Ellen Gordon and Eileen Zweig developed Operation Bookworm, a successful volunteer reading program, to improve literacy and involve more community members in their schools. They invite parents, community members, senior citizens, and high-school students into elementary schools to give K–1 students extra reading experiences. Volunteers work one on one with selected students, helping them to read. The volunteers learn special reading techniques by attending two workshops based on proven early childhood reading strategies. To further expand children's literacy, Gordon and Zweig have written a book and conduct family literacy workshops to show parents and grandparents how to get their children excited about reading with easy-to-use activities that develop and improve basic reading skills.[33]

Parents Pitch In
Teacher Teresa Jo Clemens-Brower invites parents to pitch in and help her in the classroom in a variety of ways. She asks parents to be a teacher for a day. Parents create their own plans, projects, and assessments that fit in with the themes that the class is working on in areas ranging from writing and science activities to lessons in piano, art, carpentry, French, and math. Clemens-Brower says, "Their [parents'] commitment to get time off to share something special with us showed their employers that schools and students are worthy of investment."

Parents who can't come to school help at home by preparing materials for art projects, gathering library books, and scanning the newspapers for articles of interest. Working parents appreciate the opportunity to participate in their children's education.

Experts from the community come to speak on topics suggested by parents when families come together one evening a month to learn with each other. Science projects and math games are available then for families to explore. The approaches Clemens-Brower uses build a stronger sense of community among the teacher, the students, the parents, and the community.[34]

Sidebar 9.7

Chris Horwath, Parent Volunteer Coordinator
Tyner Academy, Chattanooga, Tennessee

Parent Involvement in the High School

Parent involvement is quite different at the high-school level. For years we have been programmed into thinking that, when students get to high school, parents need to take a more hands-off approach. In reality, though, these are the years we need them to be there. Teenagers will tell us they don't want us around, but so many times actions speak louder than words. Because there are so many changes in the family structure today, the term "parents" can include grandparents, aunts, uncles, or even older siblings. At Tyner Academy, which is a science, math, and technology magnet school, we have several types of parents who volunteer in our programs. One particular story from a "parent" comes to mind to reinforce the message that involved "parents" do make a difference.

One student's grandmother volunteers once a week. She is 71 and retired but continues to stay active through different types of volunteer work. When she comes to Tyner Academy, she usually comes for the whole day, taking a break only to get some lunch in the school cafeteria. One particular day she came back to the Parent Volunteer Office, her face all aglow to tell me the following story: "I was admiring the students' artwork, which is displayed in the cafeteria, and one of the students came up to me and pointed to the one I was looking at and said that it was his picture. He asked me if I liked it. I commented to him that it looked dark. He explained that the art teacher told him to paint how he was feeling so that's what he did. I told him that it looked like he had broken up with his girlfriend. He said he did—the day before he painted the picture. I looked at him and told him I thought he looked like he was doing better. Then he turned, smiled, gave me a hug, and went on his way."

This was a win-win situation. The student won from having a sincere "parent" take time to notice, talk with, and encourage him. The parent was fulfilled when the student reached back to her.

Parent involvement in the high school is a baby-step process with each step leading to a progressive walk from parents to students and staff. This was revealed in another story told by one of the teachers. When one of our parents chaperoned an art field trip, the teacher learned that he was a carpenter. Now he is building some much-needed cabinets in the art room. The teacher is excited. Connecting with this parent has helped her and enriched the students' learning environment. Once again, we have a win-win situation.

Sidebar 9.8

Nancy Goldberg, Kindergarten Teacher
Consolidated School, Kennebunkport, Maine

Parent Volunteers Create a Win-Win-Win Situation!

Parent volunteers greatly enhance the children's educational experience. A successful volunteer program is enjoyable for the parents and students, beneficial for the children, and helpful to the kindergarten teacher.

- *How Parents Contribute.* Throughout the school year, parents volunteer daily in my kindergarten program. They work in the reading, writing, math, and science programs, computer lab, school library, on field trips, cooking, and arts and crafts projects. Several times a year, parents share particular areas of expertise and involve students by making a classroom presentation. Parents open snacks, zip coats, tie shoes, clean paint brushes, and put away materials. They listen to and encourage children. Involvement continues at home for parents who volunteer to make "shape books," mix play dough, and bake cookies.
- *Making Volunteers Comfortable.* I prepare volunteers by offering a volunteer training session early in the school year and additionally as needed. Parents are free to sign up to volunteer as often or as infrequently as they like. Some parents volunteer twice a week and others twice a year. I encourage whatever schedule works best for the volunteers. Preschool siblings are also invited into the classroom when parents volunteer so baby-sitters are not needed at home. This provides the added benefit of introducing the classroom to future students while facilitating visits from parents who otherwise could not volunteer. Stepparents, aunts, uncles, and grandparents, in addition to the students' mothers and fathers, volunteer in the classroom.
- *Benefits for Everyone.* Parent volunteers contribute greatly to the success of my kindergarten program. They allow me to provide an enriched curriculum to my students. Parents often have expressed the benefits they receive from volunteering. It provides volunteers with the opportunity to (1) view firsthand their children's routine and peer interactions, (2) learn how I teach academics and social skills at school, and thereby (3) reinforce and extend to the home the learning that begins in my classroom. Volunteering opens up parent-teacher and parent-child communications and keeps parents informed about school activities. My primary interest in soliciting parental involvement and support is in maximizing the learning experience for my students. The fact that parents and I derive real benefit from the program makes it a winner for all concerned!

Families Learn about School Practices

If parents are not involved when schools introduce new practices, they often misunderstand and oppose the changes. Many schools have realized the value of helping parents learn about the new practices. In some schools parents participate with educators on planning teams to research possible new programs, and parents are invited to give feedback to educators about proposed programs *before* they are adopted.

Sidebar 9.9

Deanna Nadeau, Looping Teacher, Grades 1 & 2
Montello School, Lewiston, Maine

Parent Workshop

At the beginning of the school year I invite the parents of my first graders to a Parent Workshop. I hold the workshop in the evening during the first month of school. This workshop was designed for multiple reasons.

First, we discuss the stages of reading and writing development and how they parallel oral language development. Children learn to talk because it is part of their everyday environment. I stress to parents that if reading and writing are part of their environment, also, then learning to read and write can happen naturally. We read and write ALL day long in my classroom. I expect my students to read, or be read to, every night. I model reading to and with children for the parents. My goal is to "hook" the adults to books, so they can take this enthusiasm home to their children. I remind them that reading competes with video games, CD players, and lots of other activities. They have to make books come alive!

Second, I want parents to have an idea of what goes on all day while they are away from their child. We "walk" through a regular day in our classroom. They become first graders in a sense. We start with Morning Meeting where we greet each other and then share something. Reading songs or poems on large chart paper follows this. We then have Reading Workshop, Story Workshop, Writing Workshop, Literature Groups, Math Workshop, a science or social studies block, and center time. I explain what happens during each of these.

I feel strongly that parents need to know what's going on in schools. My door is always open to visitors. This workshop is informational and exciting for parents. It is a way for them to stay connected. Many work full time and cannot volunteer during the school day. If they come and get ideas about what they can do at home, then they remain connected. Children don't usually remember everything that goes on in their day. At least, with this evening presentation, parents know what's happening. Throughout the school year I send home *Nadeau's Notes*, a weekly newsletter about what we did that week, ongoing themes or units of study, and upcoming events. Student writing is published in the newsletter. Communication between home and school is crucial to getting the best education.

Metro East Parent Connection in East St. Louis, Illinois
One goal of the Solid Foundation program, started by the Illinois Family Education Center in 1999–2000 is to build a partnership between families and schools and to bring change to the schools at the same time. The center has implemented six different programs in 33 schools and six districts in East St. Louis.

Reading Compacts are jointly agreed on responsibilities for the teachers, students, and parents in the area of reading. Kindergartners through third graders have activity sheets that encourage family reading and are a part of the *School-Home Reading Links*. *Reading at Home* is a course for parents to help them contribute to the development of their children's reading habits. *Family Reading Night* is an evening of reading activities for the school community. The *Family Resource Library* is set up in the schools to provide resources for the entire family, and in the *Interactive Reading Workshop*, parents and teachers develop interactive storybooks together for the library in their schools. Through these activities, parents learn how to help their children at home as well as the ways in which the schools teach reading and writing.[35]

U.S. History Class for Parents in Tucson, Arizona
Mitch Dorson, social studies teacher at Foothills High School in Tucson, Arizona, invited parents of students in his junior U.S. history class to come back in the evening to participate in a shortened version of the course their children were taking during the day. Thirty parents signed up and 25 of them attended at least one session. The course for parents covered much of the same material presented to their children although Dorson added some material he felt would be valuable for parents. Topics covered included the Constitution and the Bill of Rights, the westward movement, slavery, the Civil War and Reconstruction as well as more recent history, such as McCarthyism, the civil rights movement, and the Vietnam War.

Dorson recruited seniors who had taken his class as juniors to act as teaching associates for the evening class. The students made a variety of presentations, one of which featured a student portraying Marilyn Monroe as an introduction to the 1950s. Parents loved the course and said that it had many unexpected outcomes, one being improved relationships with their children. As one parent explained, "Armed with the information and slants you were using in class with the kids, I had specific topics to start . . . discussions. Instead of asking *what* she was studying in class, I could specifically ask her what she thought of, say, the Bisbee incident. At first she was a little brief in

Sidebar 9.10

Holly Ciotti, English Teacher
Glendale High School, Glendale, California

Sharing Writing with High School Parents

The recent hand-wringing over low student achievement has focused on parents almost as much as on teachers and students. "If parents participated more in their kids' education . . ." is a common sentence starter. But "parent participation" is a term with multiple meanings—mixing finger-paints in kindergarten, chaperoning a field trip, signing a test paper, attending open house. Yet these activities are more "teacher support" than "parent participation." Their function is not directly academic. I wondered if there were any ways that parents could participate in academic tasks, especially in a high-school English class, and whether student achievement would be affected.

Based on what students do in class—read, write, listen, discuss—I decided to ask the parents of one of my ninth grade classes to do the same. I chose the following activities: (1) write an essay on the topic "Everyone knows what it means to be the parent of a high school student . . ."; (2) join me and their teens in a goal-setting conference to set literacy goals for the year; (3) create a parent-teen poetry magazine, where they would submit their favorite poem along with a paragraph on why they like it so much; and (4) respond to literature in a literary correspondence journal about the themes in *Romeo and Juliet*. The students would write letters to their parents about the play, and the parents would write back their reactions and comments.

Although my students were overwhelmingly reluctant to involve their parents in any way ("My mom'll think this is dumb."), I persevered ("Don't underestimate your parents—you may be very surprised."). In fact, they were surprised. Each request was preceded by a letter explaining what we were doing in class and constituted an "invitation" to participate in our English curriculum. Although parent participation was not fully 100 percent, on some of the activities, it came close. The parents wrote essays, attended goal-setting conferences, sent me their favorite poetry (in other languages at times), and corresponded with their teens about the woes of Romeo and Juliet.

There was only one problem at the end of the year: The "parent participation" students did no better academically than my other classes, and on certain measures, they did worse. But, when all was said and done, maybe I had asked the wrong question. Instead of *Can parent participation improve student achievement in ninth grade English*, I should have asked, *What are the benefits of parent participation in ninth grade English?* In this light, I can count the following as clear benefits:

- The parents experienced the same writing demons their teens did. How should I start? What can I write about? What if it's no good? Parents could empathize with their teen's academic dilemmas because they had them, too.

(continues)

> **Sidebar 9.10** *(continued)*
>
> - I noticed early in the year that I was more patient with the parent participation group than with my other classes. I think that due to regular contact with parents, I felt more sympathetic to the students. Knowing them as both students and as children of their parents lent a dimension to our rapport that would not have been possible otherwise.
> - The parents experienced a positive relationship with a teacher. That is, when I called home or wrote to parents, it was not to rebuke a student or to complain; it was to include the parents in our class, to whittle away at those barriers between school and home.
> - The goal-setting conference especially was a worthwhile activity for parents and teens. Frankly, I was less concerned with whether the goals were accomplished than with the fact that parent and student spent time thinking about realistically planning for the future. Ninth graders are barraged with requirements—for graduation, for college, for career choices. It was good for at least some things to be in their control, even if it was only a decision to read a book one hour a week.
>
> Although writing and reading are often struggles for my students, I feel strongly that literacy can be shared discovery and shared pleasure.

her responses, but after a few of my classes, she was even initiating discussions, asking *me* if we talked about a certain subject of interest to her."[36]

Family Writing Workshops in Amherst, Massachusetts
Susan Connell Biggs has been inviting her students and their families to participate with her in writing workshops for the past 10 years. She does this because she wants her students' families to gain an understanding of how she teaches writing in her classroom. What better way to do this than to ask parents to be writers, too? As Biggs says, "The best way to understand the writing process is to experience it. We begin by brainstorming a list of memories."

After going through the writing process and writing a piece, parents and students gather to share what they've written. Participants are encouraged to listen carefully and thoughtfully to each other and to respond to each other's writing in nonjudgmental ways. Amazing things happen as a result of the writing and sharing during these workshops. Through the writings, parents and students speak to each other in touching and funny ways. These workshops help bridge communication barriers that so often afflict relationships between parents and their teenagers. One parent came to understand, for the first time,

> **Sidebar 9.11**
>
> *Deanna Nadeau, Looping Teacher, Grades 1 and 2*
> *Montello School, Lewiston, Maine*
>
> ### "Math with Mom"
>
> As a Mother's Day activity, many teachers have Muffins with Mom. During the first-grade year of my 1–2 loop, I invited moms in for muffins, and my students sang Mother's Day songs to their guests. I did not want to repeat this activity with the same moms in our second-grade year so I came up with "Math with Mom."
>
> Mothers were invited to spend an hour and 45 minutes with us. The time was spent doing various activities with the whole group or in small groups. In all I planned five different activities for the day.
>
> An example of a whole group activity was a lesson on probability called *Shake and Peek*. This idea comes from a math workshop I had attended years ago. There are 10 marbles in a special shake-and-peek box. Each person shakes the box and then peeks at the marble through a hole cut in the corner of the box. The whole group keeps a tally chart of all the peeks, and then the child and mom have to predict the amount of marbles of a certain color. Their guesses are recorded in their Math Journals.
>
> An example of a small-group activity was an addition game I called *Addition Around the House*. This game is played with two or more people using dice.
>
> Not all children had their moms visit that day as some are full-time working moms, as I am. I take care of this issue by asking those who have moms in attendance to share their mom with a friend. This usually works well. "Math with Mom" was very successful. The discussions I heard that day were interesting. Moms were a bit surprised by the children's and their own mathematical knowledge.

her son's perspective about the time in his life when he first got into trouble in school. Biggs says, "our stories can be gifts to each other, a means of connecting to each other in a way that is unique from any other process. When we stop sharing our stories, we stop truly knowing each other."

As parents gain insight about the writing process, they can be more understanding of what their children are experiencing and help them when they have difficulty. After attending the writing workshop, one parent with a daughter who struggled with writing would sit down with her daughter and begin to write when she noticed that the daughter was having difficulty. The daughter said that this made it easier for her to write.

Biggs offers this workshop for extra credit in her course because she understands how complicated parent-adolescent relationships can be. While parents do want to continue to be involved in their child's education when they

> **Sidebar 9.12**
>
> Deanna Nadeau, *Looping teacher, Grades 1 and 2*
> Montello School, Lewiston, Maine
>
> ### "Discovery with Dad"
>
> In our first-grade year we invited dads to "Donuts with Dad" and shared Father's Day songs with them. Our second-grade year we had "Discovery with Dad" instead.
>
> The classroom was set up in five stations. Each station had materials and an instruction sheet. The five stations were called *Everyone Is Different*, *Does It Roll?*, *Bottle Fun*, *Tomatoes & Toothpicks*, and *Sink or Float?* Dads and students were divided into small groups, which rotated to each station.
>
> *Everyone Is Different* was a fingerprint- and magnifying-glass activity. I invited an officer from our local police department to help with the actual fingerprints. Then the children and dads looked at them with magnifying glasses and discussed the differences they saw.
>
> Different objects and a ramp were used for *Does It Roll?* First, the participants made predictions; then they tested them and recorded their results.
>
> Each activity took about 15 minutes. The trick to planning this event was in choosing activities that would hold everyone's interest and take about the same amount of time. "Discovery with Dad" was a huge success. The interaction between dads and students was heartwarming.

move into high school, students, who are trying to become independent and who want to be accepted by their peer group, may not be as enthusiastic. By offering extra credit, she gets around the student's reluctance to admit in front of their peers that they may like the idea of having their parents involved.[37]

Parents Take on New Roles in Schools

> *Discovery School (a school of choice) is guided by an advisory group of four teachers, six parents, three community members, a school board member, and the school's Administrative Assistant. The group is highly committed and makes many decisions, including the annual selection process to determine who will attend the school.*
>
> —The Boyer Center, *Boyer Best Practices 2001*

Parents want to be involved in schools in different ways. Some prefer to focus their energies on what they do at home with their children, others like

Sidebar 9.13

Elaine M. Magliaro
Teacher, Malcolm Bell Middle School
Marblehead, MA

Bringing Parents and Children Together Through Poetry

Poetry is an important part of my life. It has helped me become a better thinker and writer, and it has also given me a means to express my thoughts and feelings to other people in a way I could not do as well before poetry entered my life. When I began a research project with my second-grade students, I hoped that, if children and their parents read poetry together, they might develop an appreciation for poetry, and, by sharing feelings about the poems, parents and children would come to know each other better.

First, I purchased multiple copies of single-author poetry books and anthologies. I read some poems from each book and discussed them with my students. I then left the books on a table in the classroom for children to look through and select the ones they wanted to read in class and to share with their parents at home.

After spending several days in my class reading, discussing, and doing other activities, the children took books home. Each night for a week they read poems to their parents, parents read poems to the children, and parents and children discussed the poems they had read together. In their poetry journals, the children responded to the poems they had shared in the same way they had previously done with poems we read in class.

To get feedback on the project, I asked students and parents to complete questionnaires. Some parents, though, had already responded to the poems in their children's journals. Some even wrote poems themselves. The project was a success. Children really enjoyed the poetry reading, and more than half the class said they liked to write poems. One boy, for example, wrote, "When I read poetry, that encourages me to write more poetry. Writing poetry gets my imagination going."

Parents were equally positive—18 of the 22 families returned the questionnaires. Many parents said they had learned more about their children because of the home poetry assignments. One mother noted that she had learned about her daughter's sensitivities. Another commented on her son's sense of humor when he read and laughed at funny poems. One girl's mother wrote: "She [her daughter] is more insightful about what we read and more creative in her writing than I had thought." The father of one of my most avid readers wrote: "Her grasp of the content of the subject matter was more comprehensive than I had previously thought." Sharing the poems in some homes involved the whole family. The parents' responses showed me that parents learned more about their children's gifts—and they enjoyed having the opportunity to spend time with the children in such a positive way.

I wanted to open up the world of poetry to my students and their parents. I knew I had succeeded when I saw the excitement on children's faces every time I

(continues)

> **Sidebar 9.13** *(continued)*
>
> brought a new poetry book into class, when students took poems outside to read at recess, when they sit together reading poems, when they requested more books to take home to share, and when they chose to write poetry instead of prose during writing time. The parents' responses gave me written proof of something I have believed for a long time: literature, especially poetry, not only touches the mind but also the heart—and it can open up new areas of communication between parents and children.

volunteering in classrooms or actively participating in the PTA. Successful schools offer parents diverse opportunities and do not forget to invite parents to serve on school site management teams, curriculum planning committees, and strategic planning teams, because, as Seymour Sarason, noted educational scholar, says, "the decision-making process should reflect the views of all those who will be affected by the ultimate decision."[38]

Parents Serve on Action Teams to Build Family Partnerships
Three Baltimore schools have improved family and community involvement in a variety of ways with the help of The National Network of Partnership—2000 Schools at Johns Hopkins University. Each of the three schools—Maree G. Farring Elementary School, James Mosher Elementary School, and Cross Country Elementary School—created an Action Team for School-Family-Community Partnerships. This team of parents, teachers, administrators, and community members works to maintain strong connections among schools, families, and the communities. Each member of the action team serves on one of six committees: parenting, communicating, volunteering, learning at home, decision making, and collaborating with the community.

The focus of each action team differs by school. Maree G. Farring emphasized volunteering and learning at home, for example, by assisting families in helping children with reading and math. Volunteerism helps families become familiar with the school's daily operation and curriculum. Because volunteers also share their knowledge with other families, the communication between parents and teachers increases.

James Mosher Elementary School chose volunteering, too, as well as communication. "Pops on Patrol," six grandfathers and grandmothers, patrol the school every morning and afternoon, showing students the importance of

> **Sidebar 9.14**
>
> *Karen M. Christian, Teacher*
> *Michigan*
>
> ## Sharing Teaching Philosophy with Parents
>
> You asked what I learned from working with parents. Where do I begin? Working with the whole family of the child is the most important thing a teacher can do. Parents/caregivers/ families are the most important, constant, and significant educators in any child's life. I do home visits for all my students before school starts because I realize that working with the whole family is the only way to go.
>
> I suppose I also learned that it is important to explain to parents what your teaching philosophy is and why. At least that gives parents something to go on when making a home-school connection. The back-to-basics vs. progressive educational debate is still a hot one in schools. Some parents expect their kindergartener to bring home completed phonics worksheets everyday, while others would be mortified to walk into the classroom and observe that kind of an activity occurring. With so many diverse opinions on education today, it is important to clearly state your opinion and philosophy to the parents of your students.
>
> When asked by parents how I teach kindergarten, I always talk about using themes to set up hands-on center-based learning for my students. For some parents, this is exactly what they want to hear. For others, they wonder how will their child learn to read if all he does is play. To answer the latter question and to reconfirm the former, I produced a PowerPoint presentation to share with parents. The presentation showed the children in my class interacting and working during center time. Alongside of these photographs, I explained what children were learning when they played with the materials at the various centers.
>
> I used this presentation at both parent-teacher conferences and kindergarten roundup. At parent-teacher conferences, I set the presentation up in the front lobby of the school where all parents attending conferences K–5 would be able to view it. This had two purposes: to educate parents about the kindergarten experience and to show parents how we were using technology in our building since we had just passed the largest bond in our county and taxpayers were wondering where their money was going. (A slightly more hidden agenda was that the presentation also kept the parents occupied when I was running a little behind in my schedule.) I also used the presentation at kindergarten roundup. The new parents gained a more in-depth explanation of our kindergarten expectations than did parents in the past. Also, it was of interest to the new children because my district does not offer any other way for them to see the classroom or the types of activities we have at school before they begin in the fall.

Sidebar 9.15

Janet Russell Theriault, Parent
Freeport, Maine

Parents in Partnership: Freeport Middle School

On a warm, September evening, parents, staff, and students at Freeport Middle School gathered for the school's second annual Extracurricular Fair. Conceived the year before by the Middle School's Parent Team to help parents and kids discover together what activities are available at the school, the fair would prove to be the culmination of a six-year journey shared by administrators, parents, teachers, and students.

Amid the 24 groups present, representing activities ranging from sports teams to Odyssey of the Mind, from the school musical to a group working toward reducing kids' access to tobacco, the most popular of all the booths turned out to be the one belonging to the FMS Parent Team. Dozens of parents crowded around the table, looking over sign-up sheets for volunteering opportunities such as being a "home-base caller," grade-level representative, chaperoning at a dance or on field trips, being a tutor, working with kids during morning reading, Project Adventure, and more. It was a stunning sight for those of us who had traveled that journey.

Five years earlier, when Principal Chris Toy took over the reins of the school, he brought with him the kernel of an idea for a parent structure untried before. It was the perfect time and place to give it a try, for entering the school at the same time was a sixth-grade class with a number of parents eager to be involved in the school beyond a strictly traditional "PTC" structure, as well as a core of other parents already there working to increase parent participation, an often frustrating and ultimately disappointing task.

The idea was this: Since middle schools are generally organized using a team approach—that is, groups of teachers working and planning in teams (sixth-grade team, seventh-grade team, eighth-grade team, unified arts team)—it seemed natural to suggest there be a parent team. Its membership? The entire parent body. Its leadership would be structured like a pyramid, with information flowing both up and down.

At the bottom would be a "home-based caller" for each 11-to-13-member homeroom in the school. These people would be the core of support for the activities planned as well as the method of sharing information with other parents by way of a grade-level phone tree activated each month. Next up would be a parent from each grade serving on a committee meeting weekly with the principal to help plan and organize monthly whole-school parent meetings, activities involving kids, etc. At the top would be a parent who would be the "parent leader." The choice of this person would be key, for this parent would serve as a member of the leadership team, composed of the lead teachers for each group in the school, who meet weekly to discuss a wide variety of issues. The parent would then share information, as appropriate, with the representatives from each grade. For the first time, parents were being offered the opportunity to have

(continues)

Sidebar 9.15 *(continued)*

their point of view available firsthand—as well as their support—to teachers and administrators as they worked together on the day-to-day details of life at Freeport Middle School.

As with any new idea—particularly one very different from the norm—there was both excitement and skepticism, as well as some concern that parents would be somehow coming between teachers and their students. An occasional breakdown of the phone tree setup left some parents feeling out of the loop and unsure how to access the parent structure. The Parent Team leaders met these issues as they occurred, working quietly and for the most part behind the scenes providing parent input to issues floated by them such as student rights and responsibilities, discipline in the classrooms, curriculum additions, homework quantities, parent/teacher conference formats, and more. The structure itself saw change and growth.

As a result, the Parent Team has forged a variety of partnerships within Freeport Middle School. Once a month, Student Council representatives from each grade attend the weekly Parent Team meeting, bringing issues with them. Students get immediate feedback and often support from parents for their concerns, such as overcrowded buses, school beautification, lunch and study hall issues, and more. Second, it also has reinforced and strengthened our respect for the qualities of leadership and caring we see in the students. Parents have supported kid initiatives, such as Disability Awareness Day and Spirit Assemblies. An FMS Café was created and set up by parents for each dance to give kids a place to cool down and relax during the often frantic atmosphere in the gym—and to give kids who don't care to participate in the dancing but want to be part of the night an alternative.

The team's relationship with the school's faculty has grown slowly and steadily as parents and staff have searched for a balance between parent concerns and "wish lists" and teachers' realities and professional expertise. With relief, teachers have given over the role of finding volunteers for field trips and special occasions to two parent volunteer coordinators on the team. School dances are now chaperoned almost completely by parents who in the past were not asked to perform this role. Parent-activated phone trees have conducted polls about homework and notified parents at various grade levels of important dates coming up. Teacher initiatives such as Project Adventure, Winter Survival, and peer mediation as well as trips supporting interdisciplinary education including whole-class visits to Boston have been fully supported by parents. Parents have watched with interest and participated when asked in the never-ending staff efforts to arrange and rearrange scheduling to accommodate and reflect system wide goals and student needs. We have come to appreciate the amazing energy and dedication with which FMS teachers have made the school truly kid centered.

Perhaps the most crucial partnership is the one among the parents themselves. The team has sought out interests and issues of FMS parents and looked for ways to provide education and information for them. Through the years, the Parent Team has sponsored meetings ranging from curricula information nights

(continues)

> **Sidebar 9.15** *(continued)*
>
> to parenting skills to substance abuse discussions to fun evenings including kids, parents, and staff. Attendance has varied from 25 to over 100. The group has put a strong emphasis on transition issues for families and sponsors a fifth-grade transition meeting each spring. Team members have created, along with high-school officials and parents, a transitions coffee for those with questions about high-school life. The team also has worked hard to improve two-way communications by producing a schoolwide directory, newsletters, and events postcards that are funded by the team. Further, at the request of the school committee, the group polls parents each year, providing input during prebudget discussions into suggested school and systemwide priorities. While fund raising is purposely not a top priority of the group, the team organizes and staffs a very successful magazine drive each fall that raises thousand of dollars to fund class trips and activities.
>
> Throughout these years, growth slowly came to the team itself. Most positions became shared between two parents. New positions were added and filled. And the team worked hard to publicize its efforts and goals down through the system's two elementary schools, which each boasted strong parent involvement. But parents were not knocking the doors down . . . until that September evening when the FMS Parent Team came of age.

school safety and punctuality. Mosher also supports home visits, father-son breakfasts, and Man-to-Man workshops to encourage fathers to become more involved with the school.

At Cross Country Elementary School the action team produces a monthly newsletter that includes summaries of workshops, a calendar of school events, the cafeteria menu, and messages from the principal, the action team chair, and the PTA. A Staff Volunteer Needs Form was developed to help teachers recruit volunteer help where it was most needed.[39]

Parents and Teachers Work Together To Improve Schools
Several schools engaged in parent-teacher action research, thanks to help from the Institute for Responsive Education, the Center on Families, and Boston University. Parents and teachers helped assess needs and then used their results in planning and assessment for individual students and in schoolwide educational decision making, curriculum development, and assessment.

The action research led eight schools to create a variety of programs. Anwatin and Northeast Middle Schools created PATHS (Parents and Teachers Headed for Success). In addition to installing telephones and answering machines in each classroom, there is now a parent visitor/guest lecturer program and a parent worker program. At Fairfield Court Elementary School, a team of home visitors visited parents twice a month, worked with them on home-learning activities, and provided them with community resources.

To assist families with children ages 10 to 24 months in the Ferguson-Florissant School District, of Florissant, Missouri, a team of parent educators visited the homes to provide parents with activity boxes of toys and other materials to use with their children. The mentoring program at Samuel Gompers Fine Arts Option School in Chicago gets male volunteers from the community to help students increase their self-esteem and academic success.

To increase parent involvement, the Patrick O'Hearn School established a Family Outreach team, which visited parents in their homes. The action research team at the Matthew Sherman Business and Government Preparatory School coordinated a parent involvement program that included home visits, teacher-training workshops, a parent center, and the Organization of Latino Parents (OLP).[40]

School Community Councils in Lincoln, Illinois
In Lincoln, Illinois, School Community Councils (SCC) (composed of the principal of the school, two teachers, and four parents), which meet twice a month, work to engage parents as partners with teachers to aid students in the achievement of academic and character goals. The Washington-Monroe Elementary School SCC established two academic and two character goals for all students: reading, studying, responsibility, and respect. Both teachers and parents get copies of the School Community Compact, outlining these goals and expectations. "Parents have a better understanding of their role in their children's education because the School Community Compact spells out their responsibilities clearly."

The School-Home Communication Committee, a subgroup of the SCC, improved home–school communication by initiating the practice of teachers sending folders home with the oldest child in each family on the second and fourth Monday of every month. Inside the front door of the school there is also a parent bulletin board. Another subgroup, the Parent and Teacher Education Committee, plans teacher and parent in-service training.

The SCC also promotes the idea of a community within the school through creative projects that require participation from school families and local businesses. For example, students created a literature garden, filling the plot with plants symbolic of stories they have read. As students share the work of planting, watering, and weeding, they learn lessons in science, geography, history, reading, and mathematics at the same time.[41]

Parents and Educators in Minnesota Plan School Reform
The community of Herman, Minnesota, faced severe problems when school enrollment decreased from 476 students in 1970 to 185 in 1999. State funding, allocated on a per-pupil basis, decreased sharply. Don Anderson, the superintendent of schools, contacted the Center for Small Towns at the University of Minnesota at Morris. Together they created a core work group of parents, students, teachers, and administrators to develop a plan with a "focus on the vision of the school rather than . . . poor economic forecasts." This group invited speakers to help teachers and residents learn about school reform. Ultimately, the team decided to focus on curricular innovations, including individualized learning plans and project-based learning, effective use of technology, and marketing to potential students. An overwhelming 81 percent of community members surveyed felt that they knew more about school issues after the action plan was implemented, and curriculum changes have met with great success. The situation in Herman shows how rural school development can be both successful and community based.[42]

HOME, SCHOOL, AND COMMUNITY CONNECTIONS

> *Kipps Elementary School was designed for community use—the gymnasium was built in partnership with the city's recreation department and is reserved for their use after school.*
>
> —The Boyer Center, *Boyer Best Practices 2001*

In many places parents, educators, and community members are working together to educate children, offering glimpses of the potential of the synergistic ideal proposed in this book. These collaborations take a variety of forms: Schools enrich the curriculum through community connections; schools and community organizations collaborate to help families and children in need; community organizations provide educational opportunities after school; and community organizations advocate for better schools. In addition, in the United States there is a long history of efforts to create community schools—schools that serve as the hub of community activity.

Enriched Curriculum

Schools and community groups working together can extend and expand educational opportunities for children by bringing the community into the school or taking children out into the community.

Antarctic Learning in Palatine, Illinois
Thanks to a sustained partnership with the Planetary Studies Foundation, Palatine, Illinois, located in Algonquin, Illinois, had already integrated the study of space and geology into the school's curricula. Yet a unique experience arose when Professor Paul Sipiera, a professor of geology and astronomy at William Rainey Harper College (also in Palatine), began a privately funded expedition to Antarctica to study geology. When he invited a teacher from Palatine's District 15 to go along, the program goals came to include not only the collection of meteorites and the sample ice for life-forms but also to "support distance education and learning through real-world experiments and scientific data exchanges with students in District 15 schools." The district transformed its curriculum by integrating science, math, and geography, and by supporting the travels of teacher Sharon Hooper, a middle-school science teacher. In addition to the advanced technology already in the schools, Motorola provided a satellite phone for the project.

Students in their classrooms listened as Hooper participated in satellite broadcast interviews. She and the students also conducted the same experiments and compared results over the phone, enabling students not only to duplicate the experiments but also to understand how geographical differences affect scientific results. When Hooper returned, she participated in a live question-and-answer broadcast. The Antarctica trip and experiments "really personalized science for [students]," and also presented a positive model of business, community, college, and school partnership.[43]

Retirees and Active Professional Enhance Science Education
Project RE-SEED (Retirees Enhancing Science Education through Experiments and Demonstrations) has been recruiting and training retired engineers and scientists since 1991 to be volunteers in classrooms where they help middle school science teachers. Five years ago this program expanded to also include people with a science background who are still working full time.

Working in the classroom once a week, these volunteers become a vital part of the students' learning experience. They collaborate with the teacher, helping with activities that are relevant to the curriculum. These volunteers help teachers with activity-based science and mathematics and inject their expertise into the students' critical middle school years.

RE-SEED is unique in the amount of training that volunteers receive prior to their classroom work. The volunteers are supplied with a kit of materials and a sourcebook with over 200 hands-on activities. They also attend several workshops reviewing and practicing demonstrations that focus on the physical science concepts.

Another innovative aspect of RE-SEED is its leadership program, which trains retirees with at least a year of classroom experience to conduct RE-SEED training programs for new volunteers. This mechanism enables a self-sustaining RE-SEED training site to be established, which has facilitated the expansion of the program throughout most New England states and Alabama, California, Colorado, Maryland, and Virginia. To date, RE-SEED has trained over 400 volunteers and has impacted 80,000 students in nine states offering more than 200,000 hours of support in science.

Surveys completed by volunteers, teachers, and students indicate that this program has a tremendous impact on the quality of instruction and the learning that the students achieve. One teacher wrote, "I now truly expect to understand physical science concepts. Before . . . [my volunteer] came in, I never thought I'd 'get it.' [He] is a gem." A student remarked to another volunteer, "You were great; you're a retired genius!"

The impact of RE-SEED volunteers goes far beyond improving science teaching. Many students do not understand what scientists and engineers actually do. Therefore, learning about science may not seem important to them. The presence of a professional scientist or engineer in a school helps students to understand how academic principles are used. They see firsthand that the math and science that they are working on in school is used in the "real world."[44]

Community Organizations Provide Before-, After, and Vacation-School Programs

Art groups, theaters, museums, and many other local community organizations offer a wide array of after-school and vacation programs for children of all ages—sometimes in collaboration with schools and other times on their own. Crossroads School in New York City works hard to connect their students with summer programs sponsored by a variety of local, state,

Sidebar 9.16

Kit Juniewicz
University of New England, Biddeford, Maine

Engineers Help Students Build Robots

While coordinating a university program that focused on connecting public schools with companies and individuals eager to provide support, I had the exciting opportunity to coordinate a project that had engineers, students, and teachers working hand in hand to create a robot for a national competition.

"US First Robotic Project" was the brainchild of Dean Kamon, an engineer and inventor, who had a difficult time adjusting to the demands of the typical classroom setting in both public school and in college. Yet after he left school, he went on to become very wealthy because of his inventions. He created the Robotic Project to address the needs of students who didn't "fit the mold." The project's mission statement explains the purpose: "Focus is on creating a demand for learning, capturing national attention through high-visibility media events, unique business and school partnerships, and by promoting real-life role models. The next generation must know that their knowledge of science and math can directly affect their future standard of living—we market the value, rewards, and excitement of engineering and technology."

US First holds a regional competition in Manchester, New Hampshire (where Dean Kamon lives), and also a national competition at Epcot in Florida. To take part in this project, students and teachers from South Portland (Maine) High School teamed up with engineers from what is now Fairchild Semiconductor to design and construct a robot. Eleven employees from National Semiconductor, 4 teachers, and 40 students were involved in the project. In January, representatives of the team traveled to Manchester, New Hampshire, where they picked up a kit of materials, learned what task the robot had to perform, and received very specific guidelines about what was allowed. The engineers, students and teachers worked together, brainstorming the robot design, and then divided into teams for specific tasks. Teams met together periodically demonstrating how the teams were interdependent.

The team had until the middle of March to complete their robot and ship it to Manchester. There was a celebration at the high school as the robot was crated and sent on its way. A mother of one of the student participants presented me with a box of homemade fudge and said, "My son hated school and was placed in an alternative program. This project has turned him around. He now believes that he does have some talent."

In New Hampshire the team competed with similar groups from across the country. The excitement was tremendous as three robots entered the arena and vied with each other to be most successful at the assigned task which was to get the most balls in a large central cylinder. It was great to see the many designs of robots and the various strategies used. These included some very aggressive moves, and the cheering heard was as loud as any heard at a high-school championship basketball game. Besides all the fun and learning, this team won the "rookie award."

(continues)

> **Sidebar 9.16** *(continued)*
>
> Just as important, though, was the greater understanding that developed between the National Semiconductor engineers and the South Portland High School students and teachers as they learned about each other. An engineer remarked, "Schools are a lot different than when I went. I didn't realize what teachers have to do." A teacher commented, "I learned more about the way industry works. This will be valuable to my students."

and national agencies. When schools know what is available in their communities, they can link children and parents to programs that best meet their specific interests and needs.

Camp Fire USA Community Family Clubs
Camp Fire USA, a nonprofit organization for youth development, recently added a new program called Community Family Club. The program, the result of two years of research, is open to all families, "regardless of the form the family takes in today's society," says the national chief executive officer for Camp Fire USA. Families meet at least once a month in schools, churches, corporations, or childcare settings to participate in activities. Older children are encouraged to take on leadership roles. When the club is based in a school, families, classroom teachers, and counselors all work together to plan the activities. The end result is that parents and community members become involved as partners with the school in the children's learning. The program, currently being piloted in seven states, is expected to be implemented in 41 states and the District of Columbia by summer of 2002.[45]

Chicago Public Schools and ASPIRA Join Forces
The ASPIRA Association, Inc., the only national nonprofit organization devoted solely to the education and leadership development of Puerto Rican and other Latino youth, and the Chicago School District have joined up to offer students a variety of extended learning opportunities. These include an afternoon enrichment program for 100 Hispanic students in three middle schools, 17 clubs for 600 middle- and high-school students, tutoring on weekday afternoons and Saturday mornings, and a six-week summer program for 40 middle-school students at risk of academic failure. Programs focus on science and math activities, tutoring, and leadership development. This partnership is composed of community and student volunteers and

Northwestern University, which is developing a longitudinal evaluation of program participants.[46]

East Harlem Tutorial Program: A 40-Year Success Story
Founded in 1957 by Helen Webber, the East Harlem Tutorial Program is a nationally recognized model for after-school education. With over 400 trained and supervised volunteer tutors, it serves 390 students in New York City. A professional staff of educators designs and directs the programs. Volunteers must apply, undergo reference checks, and participate in ongoing tutor training specializing in art, literature, and technology development. Teenagers receive one-on-one tutoring as well as the opportunity to tutor a younger child. Teens also are selected to work as junior counselors, college interns, and counselors-in-training.

The program also offers opportunities to parents. For instance, 31 workshops were held in 1998 on parenting, investing, computing, health/dental screening, and social work. The Mt. Sinai Health Care Center and other social service agencies partner with the tutorial program, and personal counseling is part of the offerings for children and families.[47]

Shiloh Baptist Church in Washington, D.C., Reaches Out
Shiloh Baptist Church in Washington, D.C., shows how a church and its members can serve the needs of the local community. In addition to adopting a local public elementary school, the church offers a math academy for children staffed by parish volunteers, and, together with the neighboring Washington Hebrew congregation, cosponsors a program for preadolescent boys. Through these projects, Shiloh Baptist promotes family and community involvement in support of children's learning throughout the school year.[48]

Faith Communities Join Hands in Jackson, Tennessee
Faith communities in Jackson, Tennessee, came together to support children's learning in a very tangible way. Adults from 23 churches offer tutoring to children each week. The program connects with the local school system through a homework hot line. The faith communities have expanded the partnership to gain support and participation from local businesses.[49]

Lunchtime Programs in Wilmington, Delaware
First and Central Presbyterian Church, located in downtown Wilmington, Delaware, sponsors a series of lunchtime seminars for employees of area

businesses, many of which feature the importance of being involved in children's education and the ways in which working parents can be involved with schools. Building on the seminars, the church, the school system, and the businesses involved decided to start an after-school tutoring program using the church facilities. The school system trains employees of area businesses and church members to serve as tutors.[50]

Schools and Community Organizations Help Families and Children in Need

Families in need often don't know where to turn for help because they do not know what services are available, or one agency works with a family without knowing what the other agencies could do to help. Collaboration among these agencies and the schools will ensure that children and families get the coordinated services that would best meet their needs. Some communities have taken the lead to coordinate community resources so that families have better access.

The Great Family Network in Escambia County, Florida

Based in Escambia County, Florida, the Great Family Network began when a judge in the First Judicial Court of Florida determined that 26 percent of all children in the county and 57 percent of all African American children lived in poverty. Judge John Parnham assembled a task force of 34 area leaders to address the needs of families in the area and to propose options for dealing with those needs. The task force looked at direct educational reform, but it also realized that "children cannot be expected to conquer such forces as hunger, poverty, and abuse." Thus, the group turned their attention to how local religious communities could help families.

The result was the Great Family Network, a network of 12 local churches, which would provide care teams to "enter into a mutual partnership with a special needs family and its school and develop a caring, trusting relationship with the family and its school." After six months of working together, the family and the care team reevaluate their work to determine the best course of action to take.

The network provided training for the care teams, including such topics as tutoring, working with special-needs families, and goal setting. The care teams made sure that families had school supplies and clothing and set up school conferences and meetings when needed. While it is too early to evaluate the efficacy of the program as a whole, families have responded positively to the program, and school guidance counselors continuously request additional care teams.[51]

Partnerships to Keep Students Healthy in Providence, Rhode Island
Because "educated people are healthier people and healthier students are better learners," the connection between community-based health education and service programs and school reform can be a powerful collaboration. Health reforms involving nutrition, general health services, and counseling and social work provide greater access for families in poor communities, especially in urban schools. A "healthy school" can lead to community awareness while decreasing probabilities of violence, racism, and behavioral problems in schools. A key element of this Rhode Island partnership is the collaboration of school, health services, and parents.

The Veazie Elementary School in Providence recognizes the importance of addressing health issues in schools because, while "school staff members do not create the health and social problems of their students . . . there is much that they, together with their communities, can do to address health and safety issues." Thus, teachers refer students to a family-school support network in order to ensure that all health and social needs of students at risk are met. Further, all students go through health and physical education courses. A family resource center complements these courses and is staffed by parents who are also Americorps volunteers.[52]

Community Groups Provide Services to Parents
Many schools and communities across the country have found ways to work together to serve children and their families.

The Referral and Information Network (RAIN) at the Fienberg-Fisher Elementary School in Miami Beach, Florida, created a RAIN room as a place in the school where parents help each other by planning home visits, making phone calls, and translating for new families who are uncomfortable asking for help from agencies as a result of their undocumented status in the United States and the stigma of having to ask for assistance. In Las Cruces, New Mexico, administrators use federal, state, and local resources to provide services to their families, most of whom do not have health insurance.

At New Brunswick High School in New Jersey, a partnership between the University of Medicine and Dentistry and the State Department of Education offers families mental health counseling, an infant care center at school for teen parents, a health education program, employment services, and recreational opportunities. Students help design and implement programs and workshops.

In Snohomish County, Washington, a parent facilitation project trains parents to be partners and advocates for their children's learning. The project works with parents of limited income to develop decision-making skills.[53]

Community Organizations Advocate for Better Schools

When public schools do not educate all children equally well, parent and community groups can join to advocate for change. Sometimes an outside organization approaches the schools and builds a coalition, as was the case with the Dallas Area Interfaith group. Using a community-organizing strategy developed by the Industrial Areas Foundation, this group approached churches in the Dallas neighborhood serving Roosevelt High School. As a result, church members and educators worked together for school improvement.[54] Rather than being upset when outside advocates knock on their doors, schools should be open to the possibility of finding ways to work with community groups for a common purpose.

Parents United for Responsible Education (PURE) in Chicago, Illinois
Parents United for Responsible Education (PURE) exists to build support for and enhance the quality of public education in the city of Chicago. PURE informs parents about educational issues, brings parents' views into the decision-making process, and acts as an advocate for parents in their relationships with the school administration. Formed in 1987 as a result of a 19-day teacher strike, PURE publicizes parents' points of view about key school issues through press conferences, forums, public testimony, and editorials. It also publishes a quarterly newsletter, *PENCIL*, and a Local School Council (LSC) newsletter, *PURE Tips and Updates for LSCs*; offers workshops to parents and to local school councils; and operates a hot line for parents. All services and written materials are offered in both English and Spanish.[55]

A National Campaign to Get Minority Parents Involved
The goal of a national campaign called "Success in Schools Equals Success in Life" is the increased involvement of African American and Hispanic parents

in their children's education. The coalition includes such organizations as the National Association for the Advancement of Colored People (NAACP), the People for the American Way Foundation, the Eastman Kodak Company, and the Advertising Council. To help parents recognize the impact they can have on their children's education, the campaign has used parent workshops, back-to-school rallies, and guidance sessions as well as making information about schools more available. The campaign provides more information to parents through a website and a toll-free phone number.[56]

Minority Parents' Group in Madison, Wisconsin
In Madison, Wisconsin, AHANA, a group of Asian, Hispanic, African American, and Native American parents, formed to help minority parents work more closely with the schools. Because some Madison high schools have only a 4 percent black population, parents were concerned that the teachers treated their children differently from the majority students. The group works to get minority parents more involved, for example, by visiting the school to see the child's environment and to remind the teachers that the parents are a part of the education process, meeting with teachers and guidance counselors, and sharing the results of the school meetings with their children or including the children in meetings. AHANA is now working to get the schools to provide parents with transportation for the school meetings and interpreters for parents who need them.[57]

African American Parents' Group: Glenallan Elementary School in Maryland
Angela Hansen, the mother of an African American elementary student, and Richard America and Tony Whitehead, researchers studying students at South Lakes High School in Fairfax County, Maryland, were troubled by the black-white achievement gap: White students test better, on average, than black students. Parents organized to examine what could be done to make sure minority children are not shortchanged in subtle ways.

Hansen, for instance, helped start the Glenallan African American Parents (GAAP) at her son's school. She explains what the parents accomplished: "They promoted enrichment activities, briefed parents on test-taking approaches, and helped boost the percentage of black third-graders scoring satisfactory on the Maryland School Performance Assessment Program test from 13.6 percent in 1996 to 42.9 percent last year."

After Hansen started GAAP, Glenallan Elementary hired additional talented faculty, increased teacher training, and raised expectations for all children. Expectations often differ on the basis of race, as Virginia Walden, a Washington, D.C., parent, notes: "'My personal feeling is that expectations from society, including our educational community, are not as great for middle-class black kids.'" GAAP wanted to make sure the school and parents had high expectations for their children. GAAP distributed sample tests and test preparation books to parents so they could help their children prepare for the Maryland School Performance Assessment.

When children saw how much their parents cared about their education, they put more effort into their schoolwork. The result is that students met and exceeded the higher expectations set for them. With achievement scores rising, the black-white achievement gap is decreasing.[58]

Community Schools

A full-service community school integrates the well-being of the community as a whole with the reform of schools by offering quality education and comprehensive support services at one site. The commitment to child and family development and well-being creates the school's identity as a "neighborhood hub."[59] Although community schools are not new in the United States, people are once again thinking about the commonsense notion that educating children well is a community responsibility.

The Molly Stark School in Bennington, Vermont

The Molly Stark School in Bennington, Vermont, serves students prekindergarten to grade 6, 50 percent of whom are poor. To combat some of the problems faced by students and families, the school adopted a community school model and now provides everything from 20 after-school enrichment programs to a lending library and dental service program. The school wants children and parents to have "the essentials of success." Child care is available both before and after school, and after-school programs range from "rock and roll" to "chess" and "jewelry making." Forty-five students practice their literacy skills three days a week after school in a program run by staff, volunteers, and local college students. With activities focused on lifelong learning initiatives, from play groups for toddlers to parent literacy classes, to local business owners and workers mentoring students, these programs share a common purpose: "[T]hrough affordable programs that are accessible to all students, we provide positive social interactions and improve academic achievement."

Community members contribute to the success of the school. For instance, thanks to the help of a retired dentist and a state dental-access grant, there is a dental treatment room. Immunizations, dental care, and nutrition information sessions are also part of the program at Molly Stark. And the school recognizes that empowering parents is a key to empowering students. Thus, the school collaborates with a local adult education center to offer Read-with-Me programs, a Community Leadership Institute, and high-school equivalency diploma classes for parents.

Making services for families available at school diminishes the stigma attached to asking for help elsewhere in the community. The results have been amazing: State test scores are rising and violence has decreased significantly. The Molly Stark School represents "the essence of a true community of learners."[60]

Elk Grove Unified School District, California: Twilight Family Learning Center
Originally opened as a year-round elementary school in 1993, the Twilight Family Learning Center program now operates at four large elementary schools in the district. These school-based centers—all schoolwide Title I programs—are open year-round. Emphasizing literacy development, the centers offer K–12 homework and tutoring activities, preschool classes, and a variety of classes for adults, many of whom are recent immigrants. On a typical night, the four Twilight Leaning Centers attract 800 students, parents, and preschool children.[61]

I.S. 218 in Washington Heights, New York
The before- and after-school and weekend program at I.S. 218 serves 600 children a day and 1000 parents each week. Activities include learning opportunities for students; classes for parents in literacy, citizenship, English as a Second Language, and other topics; and medical and mental health services. Partners in the program include the Children's Aid Society, Boys and Girls Clubs of America, and parent volunteers. That the program works can be seen in the school's attendance rate, at the highest in the city for comparable neighborhoods, and in students' math and reading scores, which are steadily improving.[62]

THE BASIS FOR ALL PARENT AND COMMUNITY CONNECTIONS

This chapter has presented only a sampling of the many ways in which parents, educators, and community members can work together to educate all children well. Figure 9.1 provides a framework for thinking about the different ways schools can reach out to families and community members. Each school or community needs to determine which practices are best to adopt or adapt in a particular context.

No single program or collection of programs will lead to better relationships with parents and community members, however, unless there is a foundation of trust and respect. Thus, everyone in a school must pay attention to daily interactions. How teachers, the principal, the secretary, the custodian, the paraprofessionals, the parents, the students, and visitors talk to each and treat each other sets the tone and can build—or block—the relationships needed to work together.

Building a strong, caring community takes time and commitment. Everyone must be included, valued, and respected—even when people disagree. Yet bound by a common purpose—the creation of a community home for all children—people working together can make a difference. And so they should try.

Figure 9.1

Parents' and Educators' Participation in the School as a Community for Learning

	Function	Parents' Roles	Educators' Roles
Schools and Families Support Each Other	Transmitting culture, values, beliefs	Through modeling and teaching instill family values, beliefs, and culture in children. Families may differ on the basis of family and social culture, class, race, ethnicity, sexual orientation.	Honor and support parents' teachings and help students to learn about each other. Create a class/school environment with curriculum and teaching methods that are inviting to children of all backgrounds. Serve as mediators and facilitators of different family cultures and children's individual needs; help parents think about "all children" as well as "my child."
	Teaching Family Curriculum	Through membership in places of worship and community organizations; through family excursions, trips, experiences (arts, sports, volunteering); and through family experiences (shopping, banking, family gatherings), parents teach knowledge and skills they believe are important.	Find out what children learn at home and try to incorporate that into school curriculum. Encourage children to share with each other some of what they learn at home.
	Teaching School's Curriculum	Organize time and space at home to help children complete school assignments. Help children with school assignments. Participate in children's learning (e.g., children interview parents about their experiences as part of a school assignment).	Plan homework assignments carefully to make sure there are important learning goals so that time at home is of value (i.e. avoid busy-work assignments). Think creatively about ways parents can be engaged with their children in the learning process; encourage parents to share knowledge and experiences with children.
Parents in School	Participating as Spectator or Audience	Attend school events, such as athletic contests and art exhibits.	Take extra steps to invite parents to the school and make them feel welcomed when they do. Look at school from the perspectives of different parents in order to assess whether all parents are likely to feel comfortable and valued in the school building.
	Advocating for My Child	Meet with teachers, administrators, and others to make sure the child's best interests are being served.	Act with the belief that all parents act in ways they believe are in the best interests of their children–even when parents approach the school in a challenging way. Work with parents to address their individual concerns.
	Participating in Child's Learning Team	Work with teachers and other school professionals to plan appropriate educational programs for children.	Make parents feel welcome and encourage them to participate actively. Use parent-friendly language and explain educational jargon; invite and answer questions; listen and learn from parents.
	Volunteering	Share expertise related to their work, heritage, community resources, community values or needs, and teaching and learning with students and teachers.	Value parents' input. Recognize that parents are their children's first teachers and learn from them about their children, their cultures, and the community. Recognize that many parents have opinions, ideas, and expertise about teaching and learning; engage parents in conversations about different ways of teaching.
	Learning	Learn about new practices along with teachers. Some parents may share in the planning of activities or new practices and programs.	Recognize that changing teaching and learning practices are of major concern to parents ("Will this new practice help MY child to learn better?") and require new learning not only for educators but also for parents.
	Decision Making	Advise educators, who actually use their input, which can range from feedback on surveys to open discussions on issues as well as participation on school decision-making councils.	Understand why and how educators' expert knowledge and professionalism can be used to include, not exclude, parents from the decision-making process.
	Advocating for All Children	Recognize that all parents have the responsibility of making sure that the school meets the needs of ALL children–not just MY child. In concert with other parents, advocate for programs and practices that will be of value to all children.	Recognize that parents' perspectives may differ from those of educators; invite parents to share their perspectives. Work with parent groups to plan programs that meet the needs of all children. Acknowledge that adjustments may be necessary when school programs do not work for some children.
Parents, Educators, and Community Members Work Together	Leading and/or Volunteering	Act as elected, paid, and/or volunteers or leaders in community groups that serve children: houses of worship, town/city government, social service agencies, community groups (Boy Scouts, Girl Scouts, Little League, etc.). Help to establish a working relationship between these organizations and the school. Community members volunteer in schools.	Keep in mind that parents lead very busy lives and that they serve children in many ways outside of schools. Through parents who are also community leaders, build connections with other organizations that can help the school serve children more effectively. Recruit community members to volunteer in schools. Use community resources to enrich curriculum through field trips, collaborative programs during and after school.
	Advocating for All Children	Advocate for the well-being of ALL children in the community through active participation in above groups, individually (e.g. letter writing), and with others in grassroots organizing and other groups.	Recognize that educators can and should advocate for social justice and the well-being of all children in the community. Support grassroots organizing efforts by inviting leaders into the school and finding ways to work together.

10

RETHINKING WHAT IT TAKES TO EDUCATE ALL CHILDREN

Oklahoma City
Columbine High School
John Wayne Gacy and Jeffrey Dahmer
September 11 and the Anthrax Aftermath

Our children live in dangerous, unpredictable, and confusing times. For most people, "Columbine," "Oklahoma City," and "September 11" evoke images and emotions that are at once frightening, sad, incomprehensible, and truly painful to think about. In the 1950s children worried about the possibility of nuclear war; today children must find ways to cope with fears about buildings exploding, school shootings, mass murderers, planes flying into skyscrapers, and the threat of breathing invisible, deadly bacteria.

Adults are also fearful and worried that even worse things might happen in the future. Difficult as they are to consider, however, these events raise important questions for public schools and their communities because children are always educated in a particular time and place. But these high-profile events and people represent only a part of the context for public education in the first decade of the twenty-first century.

THE CONTEXT FOR PUBLIC EDUCATION TODAY

Other aspects of children's lives in the United States today also point to the need for rethinking what it might take to educate all children well. Consider these statistics:

- The poverty rate in 2000 dropped slightly from 1999, but 31.1 million people in America were still poor. And, even though the statistics for blacks and female-householder families were at record lows, more than one in five black families and those headed by females were poor.[1]
- The number of pregnancies nationally for females ages 15 to 19 in 1997 was 90.7 out of 1,000.[2] Or, put another way, one in 10 pregnant women were teenagers, many of them unmarried.
- From 1947 to 1997 the percentage of mothers of school-age children who work outside the home has gone from 27 percent to 78 percent. As a result, an overwhelming number of children now live in households where both parents (or one in a single-parent household) work.[3]
- In response to a 1997 survey of ninth to twelfth graders, 50.8 percent admitted drinking alcohol during the 30 days preceding the survey, and 31.1 percent indicated that they had their first alcoholic drinks before the age of 13.[4]
- In 1999, 7.3 percent of white persons ages 16 to 24 were high school dropouts. The numbers were much higher for other groups: 12.6 percent of blacks and a disturbing 28.6 percent of Hispanics.[5] Or, put another way, the national high school graduation rate in 2000 (the percentage of adults 18 to 24-years old who graduated or earned a General Equivalency Diploma) was 86.5 percent, but, although the figure for whites was 91.8 percent, it was much lower for blacks (83.7 percent) and lower still for Hispanics (64.1 percent).[6] One in three black men ages 20 to 29 is under correctional supervision or control.[7] At the start of the 1990s, more black men ages 20 to 29 were controlled by the criminal justice system than the total number of black men in college.[8]
- More than 5.5 million children were receiving special education and related services in 1998–1999. This number has increased steadily since legislation was first passed in 1975. It increased 30.3 percent in the last decade alone, outpacing the rate of growth of both the population as a whole and total school enrollment.[9]

These statistics represent only a small sampling of issues that comprise the context for the challenge of educating children today. There are other concerns about public health and the environment, for example, the increase in the number of children with asthma and autism and the effects of lead poisoning on learning. We live in a world of constant change and information overload—news reports of negative events 24/7; access to too much information online, in print, and on the airwaves with no time to process or evaluate it; and a proliferation of electronic tools we haven't learned to wisely use. Cell phones ring during church services, and, thanks to these and portable computers, people can take their work with them to the beach or the mountains. For many, the eight-hour workday is a relic of the past. Technological advances and research in a variety of fields create new ethical problems for us to consider before we have even resolved the ones that came before. Who is the legal parent of a child born of a surrogate mother? Should children's difficulties in school be treated with medication before other alternatives are tried? Now that researchers have actually cloned a human embryo, should new laws limit further research in this area?

The destruction of the World Trade Center and part of the Pentagon on September 11, 2001, made the world even smaller and more threatening than it ever seemed to be before. If some people once thought the United States could go it alone, even they must now know that isolationism is not a possibility in the twenty-first century. And these events also have raised more questions about how a democratic society can preserve freedom and protect its citizens at the same time it values personal privacy, diversity, and tolerance for the views of those with whom we disagree.

Because the context for education now is much different from what it was in the past, we cannot educate today's children for tomorrow's world in yesterday's public schools. And in such complex and challenging times when so many factors affect the education of children, we cannot expect the schools to do the job well without more help.

Not only do we need the creativity, imagination, and commitment of all citizens to meet the current challenge, we also may have to give up the idea of thinking that management plans and new laws will ever be enough to solve such complex problems. The quick fixes we have embraced in the past—new programs for every problem—are like Band-Aids on a wound: They may make us feel better for a time, but they do nothing to heal the wound. And if the Band-Aid is not sterile, a tiny cut can become a major infection.

Because we have too often addressed symptoms of problems rather than their causes, we continue trying to solve the same problems many times. Futurist Alvin Toffler notes that "[s]ystematic research can teach us much. But in the end we must embrace—not dismiss—paradox and contradiction, hunch, imagination, and daring (though tentative) new synthesis."[10] Improving public education must be a continuous process of inquiry. Parents can always find some response for the child riding in the car who asks, "Are we there yet?" But because the journey to improving education will never be finished, the answer for adults who ask this question about school improvement must be "We will never be there!" The "doers" among us will have to find ways to get comfortable with uncertainty and unfinished tasks.

Because the best innovative ideas may come from unexpected places, both doers and thinkers are needed. We need to get as many different people as possible to participate in the process. When people bring many diverse perspectives and self-interests to the table to reach agreement, the result may be conflict. On the other hand, when the purpose of the discussion is discovery—finding new ways to look at old problems—these multiple perspectives can enhance creativity. In fact, the ideal paradigm of the home-school-community relationship proposed in this book is less a model than a process.

THE IDEAL PARADIGM: ASSUMPTIONS

Several assumptions, developed in more detail in previous chapters, underlie the need for reconceptualizing the home, school, and community relationship. These statements can be viewed both as the basis for the new paradigm we propose and as arguments in its favor.

Education is more than schooling. Our rapidly-changing society and the increasing complexity of the challenges we all face mean that neither parents nor teachers by themselves can educate all children well. Everyone in the community must help.

Although children are schooled for only a few hours a day, they are being educated all of the time—in school, at home, in the community. When educational philosopher Jane Roland Martin argued for conceptualizing the schoolhouse as a schoolhome, she envisioned a place where all children were valued and accepted, where personal and social skills were as important as academic knowledge. Certainly children need schools that care, but they also need communities that care—what we call a "community home."

Students' personal growth and social development are as important as academic knowledge. All children can benefit from living and learning in a community home.

In order to make their ways in the world of the twenty-first century, our children will need a rigorous education that provides them with the ability to think critically and to solve the many unstructured problems they will face. However, they also must learn how to be caring people who can work well with others in the workplace, get along with their neighbors, maintain homes and families, and parent their own children. Children once gained this other half of education from parents in their homes, and most children still do. But other children live in families that cannot provide what they need. Parents may be too busy, or they may be unable to teach what they never learned.

Even though this knowledge and these skills are critical for emotionally healthy adults and essential to society as a whole, they seem not to matter to those who set the agenda for public schools today. If, as some theorists argue, we measure what we value, then the proliferation of standardized tests and learning standards in states and schools across the nation clearly suggests that only academics matter. Yet what kind of a future are we creating if our children learn to read and write well but become parents who do not read aloud to their own children and have no love for words to share with them? Or who quickly and accurately solve mathematical problems but, as adults, do not know how to resolve a disagreement with a coworker, spouse, or teenage child?

If we value what Martin calls the 3 Cs (care, concern, and connection) as well as the 3 Rs, we should consider additional ways to measure students' learning and the schools' ability to teach by example. For instance, in addition to reporting the results of standardized tests, we also could measure such factors as school climate and parent and student satisfaction and tout these results as publicly as a school's test scores.[11]

Parents' focus on "my child" and educators' focus on "all children" must be extended and reconceptualized to a community concern and commitment for educating all children as "our" children.

The ideal of a community home for all children begins with the recognition that *all* children deserve no less than what only *some* children now have. Middle-class parents, who are educated and have financial resources that other families lack, provide their children with a variety of educational opportunities outside of school, such as music lessons, summer camps, and computers at home. But they are also the ones who have the political savvy to influence what the schools do for their children. Thus, when budgets are tight and cuts must be made, these parents work together to make sure that school boards do not make decisions that affect their children negatively.

When such issues get debated and decided, the voices of other parents are missing because they do not speak English well or do not understand the politics of school. As educational historians David Tyack and Elisabeth Hansot note, "Amid the competition of constituencies for attention, it has been difficult to remember that public schools exist to serve all children, not simply those with the loudest or most recent advocates."[12] They argue for remembering the democratic ideal as it applies to public schools: "[T]hat is precisely one of the arguments of *public* education: discourse and action concerning public schools provide an opportunity for citizens to become concerned not simply about what is good for themselves or their own children but also what is necessary to bring about a more effective and just society."[13]

The community culture must be characterized by inquiry and continuous learning.

Since no one knows everything or has all the answers, everyone needs to work together to find better ways to educate children. And everyone has knowledge to contribute to this ongoing process. For instance, parents know their own children better than anyone else, but teachers have special knowledge about teaching and learning, and potential employers know what students will need to find jobs and succeed in the workplace when they finish school.

Combining the knowledge of everyone will have synergistic effects: Decisions made when everyone contributes will likely be better than those made by any individual or group alone. The value of this holistic, inclusive view of school improvement can be seen in the familiar poem, " The Blind Men and the Elephant," by Geoffrey Saxe:

> It was six men of Indostan
> To learning much inclined,
> Who went to see the Elephant
> (Though all of them were blind),

That each by observation
Might satisfy his mind

The First approached the Elephant,
And happening to fall
Against his broad and sturdy side,
At once began to bawl:
"God bless me! but the Elephant
Is very like a wall!"

The Second, feeling of the tusk,
Cried, "Ho! what have we here
So very round and smooth and sharp?
To me 'tis mighty clear
This wonder of an Elephant
Is very like a spear!"

The Third approached the animal,
And happening to take
The squirming trunk within his hands,
Thus boldly up and spake:
"I see," quoth he, "the Elephant
Is very like a snake!"

The Fourth reached out an eager hand,
And felt about the knee.
"What most this wondrous beast is like
Is mighty plain," quoth he;
"'Tis clear enough the Elephant
Is very like a tree!"

The Fifth, who chanced to touch the ear,
Said: "E'en the blindest man
Can tell what this resembles most;
Deny the fact who can
This marvel of an Elephant
Is very like a fan!

> The Sixth no sooner had begun
> About the beast to grope,
> Then, seizing on the swinging tail
> That fell within his scope,
> "I see," quoth he, "the Elephant
> Is very like a rope!"
>
> And so these men of Indostan
> Disputed loud and long,
> Each in his own opinion
> Exceeding stiff and strong,
> Though each was partly in the right,
> And all were in the wrong![14]

If the blind men had shared what each knew about the elephant instead of arguing about their different perspectives, they could have combined their individual knowledge to construct a more accurate understanding of the animal. The results of the process to improve schools in ways that serve all children will be more effective when people bring together their individual knowledge to create a richer pool of shared knowledge.

There are other reasons for creating communities of learners both inside and outside the school. First, students are not the only learners in the community. One study, for instance, reported that nearly half of adults 18 and above participated in some learning activities in the previous 12 months.[15] As people need to acquire new skills to cope with ongoing change in our society, everyone must be a lifelong learner.

Second, teachers are not the only ones who do or can teach children. Parents are children's first and most important teachers. But parents and community members also volunteer to share their knowledge with students in school or when students do service-learning projects or internships in the community. Adults who lead Scout troops, 4-H groups, or neighborhood and religious youth groups are also teachers. Thus, educators need to view their roles as professionals in less traditional ways. They can use their expert knowledge to guide parents and community members rather than dictating to them and can invite noneducators to teach them as well. Tyack and Hansot explain why, in their professional roles, educators need to work with others in the community:

> School leaders face the challenge of weighing professional knowledge and skill together with the need to involve community members in decisions.

> Here again there have been sharp swings of the pendulum among policy advocates from those who argue that educators are experts needing to be shielded from outside meddling to those who declare that professional wisdom is a sham. Sensible policy . . . recognizes the tension of the claims. Educators know their work better than anyone else and need not shuffle, but effective discharge of their duties requires them to work in partnership with parents and patrons. In no other way can they secure the community of interest that public education demands.[16]

Creating new ways for educators, parents, and community members to interact with each other will take time, but it can start with people who are unafraid to admit that they do not have all the answers and who are also willing to work with others to analyze problems and try out various approaches for solving them. In time, the community, composed of many individuals who see themselves as lifelong learners, can develop a culture characterized by collaboration and ongoing inquiry that is essential for the continuous improvement of children's education.

Educating all children well requires that we begin early—before children are born.

The research is clear that the first years of children's lives are the most important, not only for their intellectual development but also for their personal and social growth. Yet many young parents lack the knowledge they need to provide their children with a good foundation.

Where would they learn how to parent their children well if they themselves had no role models for doing so? As Jane Roland Martin has argued, schools prepare students to be future citizens and workers but do little or nothing to prepare them for their roles as parents. Too little has been done outside the school, as well. Scholars Sharon Kagan and Lynda Hallmark describe the current problem: "In the U.S., not only has early childhood never been a national priority, but decades of episodic, on-again, off-again efforts have yielded a set of uncoordinated programs and insufficient investment in the infrastructure. Often, the most important components of high-quality education and care—financing, curriculum development, and teacher education—are neglected."[17]

Changes eventually can be made in schools to address the problem of the forgotten curriculum, of course, but, in the meantime, communities need to consider how they can assist the young parents who need help now. Kagan and Hallmark point to Sweden as an example:

But when all is said and done, the major lesson for the U.S. to take from Sweden is a fidelity to values that cherish children and that acknowledge the special nature of childhood. We need to understand that childhood is a unique time of life, precious unto itself. We need to understand that nourishing it fully is not inimical to, but supportive of, society's best interests. As this great nation preoccupies itself with visions of the 21st century and beyond, the U.S. could make no better contribution to its own destiny than to genuinely examine its social construction of childhood and to reconsider society's obligation to its children.[18]

Providing all young children with a good start means spending public funds to provide better child care, expand Head Start programs, and, perhaps most important, raise the pay and the status of those who care for and educate the youngest of our children.

Communities need to ask themselves: "What are we doing to raise all of our children well?" And they must respond to those who reply that, given all the other programs and services that need government funding, early childhood education is not an important priority: "What will the consequences be if we do nothing?" One obvious result is that the prison population is likely to grow, and, as many have said, it costs more to support one inmate for a year than it does to send a student to Harvard. The point is that tax dollars will be spent. Do we want to spend them now to improve the lives of young children? Or later?

Schools, community organizations, and employers must work together to help parents help their children.

Some parents have very basic needs they cannot meet—they may be poor, jobless, homeless. Because of their personal struggles, they do not have time or energy to supervise children and their education more closely. Some scholars describe these barriers as an inequity: "[N]ationwide low-income parents are significantly more likely than middle- and upper-income parents to lack the paid leave and flexibility they need to help children who are doing poorly academically and children who have frequent behavioral problems . . ."[19] Yet even middle-class parents, who have the resources to provide food, clothing, and shelter, would benefit from more flexible policies that would allow them the time to participate more often or more directly.

Once again, however, we haven't found the best ways to provide the support parents need to be better partners in the education of their children. But the need is great, and the gap between privileged parents and poor parents grows wider all the time. Researchers S. Jody Heymann and Allison Earle argue that "[w]e know that parental involvement is critical. If we care about the future of all children, we need to find a way to lower the barriers parents face to taking time off from work to help address their children's challenges."[20]

We have a shared responsibility for children's education and well-being. Education writer Anne C. Lewis quotes from a recent report from the National Research Council on the science of early childhood development: "The time has come to stop blaming parents, communities, business, and government and to shape a national agenda to ensure both a rewarding childhood and a promising future for all children." Then she advises: "A dialogue—in schools, in communities, and even at broader levels—requires simply a few people to start talking about that agenda."[21]

Parents and schools need support, not criticism—labeling parents as "incompetent" or schools as "failing" will hinder rather than help them become more successful.

Labeling students as "slow learners," "at risk," or "discipline problems" is not helpful to the children we want to help become confident, contributing adults. Similarly, attaching negative labels to parents or schools that, for a variety of reasons, struggle without success will not help them improve. A collaborative problem-solving approach would work better—in classrooms, schools, and communities—than resorting to threats, penalties, and punishments. When students behave inappropriately in a classroom or a school fails to teach a large percentage of its students, the students and the school both have problems to solve. If they knew how to make things better by themselves, they probably would have done so already. Both need assistance, not condemnation, and they may be able to do with the help of others what they cannot do by themselves. Thus, these problems are not theirs but ours.

When parents, children, and schools need assistance from others, a collaborative problem-solving approach will make them more willing to make changes. Labels, penalties, and punishments, on the other hand, may decrease their motivation and increase their hostility and resentment of people in power. A brief example illustrates the difference in these approaches. The student whose teacher assigns a two-hour detention or humiliates him in front of the class will become more hostile, more resentful of authority, and turned off to school. On the other hand, if the teacher instead asks, "How can you and I work together to prevent this situation from happening again?" and assumes a share of the responsibility for resolving the problem, the same student is likely to respond positively and express a willingness to cooperate in finding a solution.

When the focus is on collaborative problem solving and continuous improvement, people will get better at what they do—no matter what their roles. And any gains they make, however small, should be valued. Extending a helping hand to those who need it will always get better results than criticizing and making their failures public.

The fact that people will always have different perspectives, values, and beliefs can be a benefit, not a barrier.

Conflict and controversy provide opportunities for dialogue and action. As the stories of both South Wellington and Springfield in chapters three and four showed, conflicts, although painful for everyone involved, eventually resulted in people working together in better ways for the benefit of children. Discussions about race at Crossroads School helped build more intimate and trusting relationships, as chapter eight illustrated.

The issues generated—or exacerbated—by people's differing perspectives are difficult, as British researchers Janet Atkin and John Bastiani, who have studied parent/school relations, recognize: "The view of home/school relations embodied in our work is not of a cozy, idealized world where everything in the garden is lovely. Whilst there are currently many interesting challenges and opportunities, there are also many actual and potential pitfalls and obstacles to progress. It doesn't do anybody any good to sweep these under the carpet."[22]

Controversies about school or community issues—if people keep working to resolve them—help everyone develop new understandings. They can be, in fact, a powerful way to learn: What one learns when emotions are high will never be forgotten. On the other hand, if a situation is never really resolved or if some people are left out of the final resolution, hard feelings will linger and build, creating a barrier to the resolution of future disagreements. Conflicts should be confronted directly, or they will fester underground only to erupt later. Thus, school and community leaders must create a process that ensures that everyone's voice will be heard and everyone will have a share in deciding the outcome.

Conflicts about education can be viewed as the working out of a public philosophy of education, which every community must do—explicitly or implicitly. As Tyack and Hansot note, this process of making decisions about education must be inclusive: "A broad-based public philosophy of education may be regarded as a community of commitment linking the people and their schools, articulated by leaders but giving citizens a strong voice. We believe that a central task of education today is to reformulate the common purposes of public education in a manner tough-minded enough to encourage controversy and broad enough to foster pluralism."[23]

When discussions—and especially disagreements—occur, people need to understand the perspectives of others and really listen to those with whom they disagree. When finding common ground seems elusive, they can join together for a common purpose of finding ways to better educate all children. Given time and commitment to this common purpose, they may begin to find or build some common ground.

We need to create new democratic forms that celebrate diversity even as we work for a common purpose.

Engaging the communities in discussions about or participation in public education today is, without doubt, a very complex and challenging undertaking. The U.S. population grows more diverse every year. Those who are white, for instance, are already a minority rather than the majority in some areas. People have joined forces to gain more power for groups that had been at the margins of society in previous decades—as advocates for women, African Americans, Latinos, gays/lesbians, to name just a few. These categories, however, are not separate but overlapping because one can be both black and a female or Latino and gay. Individuals have different shared values on the basis of their socioeconomic status or educational background. And people have other shared beliefs, which are represented by their membership in particular religions or political parties. Finally, everyone also belongs to a variety of other smaller groups—social clubs, neighborhood coffee klatches, informal networks of friends and families.

As a result of all of these connections, people's beliefs about what schools are for and how they should be are influenced, modified, or reaffirmed by their interactions with others in various settings. Given all the differences in opinions possible, one can easily see that getting agreement is likely to be difficult. Tyack and Hansot offer one basis for beginning to think about finding common ground: "What might be some common ground for agreement on such a public philosophy of education? A commitment to a common school starts with values that are not subject to empirical demonstration—in short, they are beliefs about what sort of society America should become. That is really what most discourse is about in education: a preferred future expressed as a particular kind of training for the young."[24]

Community discussions about a "preferred future" for our children can point the way to rethinking what it might take to educate all of our children well. Making any change in the status quo to accomplish what people agree should be done, however, will take place in a context of politics and difference.

POLITICS, POWER, AND THE BUREAUCRATIC BEAST

Bringing about change in any community involves politics and power relations. And change will generate conflict, especially if we continue to focus on competition rather than collaboration, on power acquisition rather than power sharing. Not only do we need to rethink our ideas about politics and power, we must avoid getting important changes entangled in the bureaucratic structures that lay like traps in the nation's capital and communities everywhere.

Politics in America has always been about winning—and losing. Even when political campaigns first begin, reporters characterize every nuance of a speech or public appearance as helping or hurting a candidate's chances of winning—even when, as in a presidential election, the voting is years away. Of course, when the votes are counted, there are winners and losers, the politicians themselves as well as all those who supported them for public office or favored one side of a referendum question over another. But is it possible to reframe the process in a way that minimizes talk about winning and losing? Could news reports, candidates for office, and advocates of ballot questions focus instead on telling citizens how their positions will affect our future as a society?

Creating a community home for all children requires multiple constituencies to work *with* rather than *against* each other. Competition is not helpful. Those who are newly elected to public office soon find out that compromise is the name of the game, but the process does not have to be one of political trade-offs. Congressional committees, as they deliberate new legislation, can work together, blending their different perspectives to create something better than the outcomes of the original proposals from either side.

Competition can be healthy, but shifting the focus to a collaborative problem-solving approach for the issues raised, with much less attention paid to who wins and who loses, could make everyone feel like a winner. Businesspeople call negotiations framed this way as "win-win" propositions. Politicians, policymakers, and news reporters should follow their lead—especially when it comes to making important decisions about children's futures.

The desire for power fuels competitions between competing interests, and this quest for power is problematic. Because we think of power as a limited quantity, then it follows that, if someone gains power, someone else must lose it. Power, however, comes in many forms. When we talk about people in power, we usually refer only to those who have what scholars refer to as "positional power," the authority to make decisions as a result of being the owner of a company, a senator, or a person appointed to hold a management position.

Less discussed, but clearly more important, is what scholars term "personal power," the charisma or leadership ability certain people have to influence others, whether they hold a recognized position or not. Principals have been fired and politicians elected thanks to the persuasive efforts of people in the community whose opinions others trust. Of course, like John Fitzgerald Kennedy or Ronald Reagan, the same individual can possess both positional and personal power. But others, whom the media never mention, can sway people's opinions by using a vast network of informal, often invisible, interactions. The fact that power comes in at least these two flavors suggests that it may not be as limited a quantity as many people assume.

The idea of "empowering people," which many argue is critical to the success of an organization, can be another way of increasing power. When decisions are made about issues of importance to them, people want to be heard. Although most parents and community members have no desire to take on leadership roles, they want those in charge to understand their concerns and take them into account when decisions are made. Even though most parents do not want to serve as decision makers in the school, they do not want schools to make changes that will impact their children in negative ways. Thus, empowering people doesn't have to mean creating positions for more people. Rather, it suggests the need to design multiple ways for people to participate in the decision-making process.

People might feel more empowered if schools and communities invited them to share their concerns about particular issues in a variety of ways, not just at public meetings: opinion surveys, straw ballots, in-home interviews (perhaps conducted by students and designed as a learning opportunity, a service-learning component for course work in school), focus groups, coffee klatches in neighborhoods—especially in those with large percentages of minorities, immigrants, and poor families, the groups most often left out of such discussions.

Public officials tend to conduct school board or city council meetings in ways that limit public comment. Individuals may be afraid to speak when the cable TV cameras are rolling—or the officials themselves may send a silent message to those who do speak to "hurry up and finish so we can get on with the real business." Board and council members, of course, don't want to stay at meetings that drag on past midnight, as happens when citizens take the microphone to argue their positions on a controversial issue. Officials then need to do some creative thinking about options other than an open mike for finding out what their constituents think and for letting them know that their views matter. When people in the community or parents in a school believe that those in charge care about their best interests and those of their children, they, too, have power.

Even though power may be more abundant than it sometimes appears, resources for solving problems are not. The process of deciding how to spend public funds frequently becomes a battle of competing interests. Most changes in schools and communities require increases in spending or shifts in the ways resources have been allocated. Once again, votes on school and city budgets are characterized as wins or losses for particular interests. The reality is that financial resources are limited. Citizens—especially as the percentage of those without children in schools continues to grow—become less and less willing to support increases in school funding. Perhaps this is a good thing: Schools may have many programs that aren't meeting the needs of the children they are supposed to help.

One way to reduce the need for more dollars is to figure ways to spend current funds differently, to "do more with less," as some phrase it. How could this be accomplished? Consider just one example from many others that could be cited. Many communities have created alternative programs for adolescents who just can't seem to function well in the regular public middle and high schools. Many—if not most—of these programs combine work experience with academic learning. Classes tend to be smaller than regular classes, and teachers create a "familylike atmosphere" in which students feel valued and cared for. The course work is individualized and personalized so students' programs meet their differing needs and interests. These characteristics echo the curriculum combination of the 3 Rs along with the 3 Cs that Jane Roland Martin proposes. But if teachers in the regular public schools conducted classes differently, in ways that were more personal and relevant for students, would the alternative programs be needed? The point here is that changing the traditional approaches to schooling could, in fact, reduce the need for special programs. The funds previously spent for alternative and special programs could then be spent for other purposes.

The typical American public high school today contains such a plethora of programs and options for students that it has been called "the shopping mall high school." Students pick and choose as if they were going in and out of department stores and boutiques in a mall, which is not a good way to educate teens.[25] Although Tyack and Hansot do not speak specifically of high schools, they suggest what communities and schools need to do when funds for education are tight:

> Superintendents face those hard choices of declining enrollments and cutbacks as a nightmare of contending forces and vested interests. In making budget cuts, it will be tempting to find targets of least resistance rather than to make decisions based on collaborative reappraisal of what makes a coherent and effective system of instruction. But defined in another way, the need to decide what is essential—and to enlist colleagues and community in that debate—can remedy the incoherence produced by easy money and rapid growth.[26]

As community and school officials tackle the question of what is essential for educating all children well and make decisions about the reallocation of human and monetary resources, they must include everyone in the process. Tyack and Hansot note that the "the new debate over purpose must recognize new conditions, diversity of interests and cultures, and the need for broad participation."[27] Because with participation comes empowerment, everyone gets to share the power, and the synergistic effects of collaboration will benefit children.

Failure to involve the whole community can thwart the success of any change, but other pitfalls exist, as well.

The "quick fix" and bandwagons are traps to avoid. Americans tend to look for quick and easy answers in the forms of new programs or practices that will immediately address particular problems. When one school adopts a new approach, the word spreads, and soon everyone jumps on the bandwagon. Too often, there is only anecdotal evidence to support such changes, not solid research. And, just as often, a few people generated the idea without carefully thinking through either its underlying philosophical assumptions or the possible unintended consequences of implementing the change.

Critical inquiry and continuous action research are essential if schools are to make real improvement in schools. Atkin and Bastiani explain: "Thinking and practice need to develop together—neither can develop in a healthy, long-term way without the other. This precludes, on the one hand, the uncritical introduction of gimmicky practices and, on the other, deep philosophical agonizing, which ignores the need to identify effective forms of action."[28] People need to avoid trying to apply a Band-Aid, which may hide the problem or address only a symptom rather than its cause, what Atkin and Bastiani refer to as a "gimmicky practice," and must not be too quick to follow the lead of other schools and adopt something new because everyone else says it's great.

In bureaucratic organizations, such as schools and community-based organizations, when power is threatened and resources are limited, those in power may hunker down to protect their turf. And any change can be a threat to those who want to maintain their positions as they exist. In an effort to maintain the status quo, some will say publicly that they support a change because they want to do "what is best for children." Their rhetoric, however, often hides a very different reality. Because they fear the loss of resources and control and the status and perks that come with their positions, they actually act in ways that undermine the success of a change.

On the other hand, others with good intentions also can prevent the successful implementation of a change. If people have spent significant portions of their professional lives developing programs and protecting them from budget cuts, they may not see the change objectively. Their intense belief in the value of their own programs can prevent them from considering alternative ways of accomplishing the same goals.

Knowing that changing practices and programs in single organizations is so difficult, many people are reluctant to consider crossing the organizational boundaries that protect individual schools and community organizations. Not only do norms and cultures differ, but everyone also has learned to guard fiercely hard-won resources. Thus, those in power will need a great deal of help and encouragement to imagine how they can collaborate and not come out losers in the turf wars.

Bureaucracy creates other traps to avoid. Linear thinking and bureaucratic processes make positive relationships among parents, educators, and community members more difficult to build. Both are also barriers to changing schools successfully. Linear thinking works against successful change by preventing people from using lateral thinking, or "thinking outside the box," to generate the kind of creativity needed to find new ways to solve very old problems. When relationships are fragile or tense, people may not trust educators to make changes that will not harm their children.

But even when people have been creative and careful to consider and research an innovation and have the support of parents and community members, it may fail. The expected positive outcomes never materialize if the innovation gets trapped in the jaws of the "Bureaucratic Beast." The change, although it might really have solved a problem, can be constrained by bureaucratic rules and procedures. Like forcing new wine into old bottles, the effort sours.

At first, the new program may work quite well, but then administrators want more accountability. Or a problem occurs that they wish to prevent in the future. So they institute new rules and procedures—reports to write, forms to fill out, cumbersome processes for requesting permission. Not only do these bureaucratic requirements constrain those in charge of the program from making decisions, they sap their energy and enthusiasm. Over time, the bureaucratic approach, although intended to make schools run smoothly and predictably, can destroy what once was a very promising innovation.

On the other hand, the Bureaucratic Beast does not have to get the upper hand. For instance, some principals long ago recognized the value of encouraging teachers to try out new ideas with more freedom than teachers typically have. These little "action research," or pilot projects, need time and space to develop before they are critiqued or constrained by administrative rules or thrown out. Positive results don't come always quickly. One principal likened these pilot projects to the bubbles created in a teapot as the water begins to boil. As more and more bubbles appear, a rolling boil emerges—and, when more and more pilot projects take hold, change on a larger scale also becomes possible. More schools should follow the lead some states have taken by waiving state rules and regulations for people to create charter schools or try out innovative approaches within existing public schools.

Giving people new freedom and trusting them to use it well can lessen the negative power of the Bureaucratic Beast and thereby foster the discovery of new approaches that will educate children better in the future than has been the case in the past.

THE IDEAL PARADIGM: FIRST STEPS

The challenge of educating all children well today is too difficult and complex for parents or educators to accomplish alone—and too important for the nation's future to waste scarce resources, by duplicating efforts or working at cross purposes. Home, school, and community cannot continue to function as separate and independent satellites, disconnected from or in competition with each other. Increasingly, we see evidence of our interdependence as individuals, as communities, as countries. And as Alvin Toffler points out, "We cannot cram the embryonic world of tomorrow in yesterday's conventional cubbyholes."[29] And why would we not want to tear down the old walls when the synergy generated by parents, educators, and community members all joining together promises so much more for our children's futures and our own?

The synergistic paradigm of the home-school-community relationship is, of course, an ideal, one that will probably never be fully realized. Yet if we keep this ideal in mind as a goal, it can guide the actions we undertake to improve the lives of all children. This ideal cannot—and should not—become a prescription. The people in each neighborhood or community must decide how they can better serve the needs of their own children. The common purpose of improving the education of children can be accomplished in multiple ways. When communities make different decisions, they will be able to share what they learn with others, thereby increasing the knowledge base from which everyone can draw.

Where should one begin?

The first step for any school or community is asking the question: "What do *our* children need?" This shift in thinking is essential—not just "my" child, "my" students, "those" kids, but "our" children. Even though self interests will never be erased, common interest can become an equal priority.

Second, focus on improving relationships between parents and educators. Parents' vested interest in the well-being of their own children is great so they are likely to be more willing participants. Yet because parents are also community members, they can link the school to other community members and resources. Parent leaders can function as liaisons between the school and the community. The process educators develop to build trusting and respectful relationships with the parents of their students will be practice for doing it with others outside the school. Yet at the same time educators connect with parents, they should not ignore the larger community because, as scholars David Tyack and Elisabeth Hansot note:

> One place to begin in creating a community of commitment is with parents of school-aged children . . . [but] one cannot build a constituency for public education solely on a coalition of parents and educators. As the population as a whole ages and the proportion of children decreases, the percentage of voters who are parents of school-aged children will drop sharply. . . . The best case for public education has always been that it is a common good: that everyone, ultimately, has a stake in education.[30]

People should practice active listening and recognize that every interaction with another person has the potential for learning something one didn't know. Active listening is helpful in two ways. First, it makes the other person feel valued and respected. Second, the possibility of finding a way to resolve differences is greater when both parties first understand the other's perspective. Because individuals have had different experiences, even a brief, casual conversation with someone can yield important information or insight.

Start small on issues that matter most to people. Because people come to any task with their own self-interests and perspectives, small groups will function better than large groups. People are more likely to participate if the task concerns something they really care about and when they know that their involvement will help determine the outcome. Working in small groups gives individuals a chance to get to know each other.

People are less suspicious of each other when their interactions are personal rather than public. Informal settings and small groups—not crowded public meetings—enhance the process of building trust, the essential foundation for any successful, continuing relationships. After a representative group of stakeholders tackles a problem, and some alternatives are clear, then it is time to get more people involved.

Everyone should be included. Even if the initial group is small, leaders should make sure that it is representative of the whole school or community. They should make special efforts to invite some individuals who are not white, well educated, and middle class. Working toward the ideal of seamless and synergistic connections requires a long process of erasing the margins that have kept some groups at the edges of the community. And it is important to remember that every person who participates will be a link to many others with similar perspectives who were not there.

Consider multiple entry points—not all of them formal. As a project continues, leaders should think of many ways that others can be involved. Large, public meetings may not be the best way to build support because people's opinions are more often influenced by conversations with neighbors or coworkers than by public presentations. Researchers Carol Merz and Gail Furman suggest that "[r]etaining and making legitimate the informal behavior that has served us well in the past may be the greatest challenge for the future."[31] They point out that allowing "ourselves to use informal means to solve problems, to allow small groups of people to seek and develop their own goals and means within a very broad organizational limits will be difficult."[32] But just because giving people freedom to discover uncommon ways to address a common purpose is difficult does not mean we should not try to do so. As history shows, the easy way is not necessarily best way.

View multiple options as possibilities with promise rather than problems. One of the greatest barriers to children's learning is that in too many classrooms every student is expected to do the same assignment in the same way at the same time. Parents and teachers know that children differ in a variety of ways, but they still can't resist the compulsion for uniformity. People also should be more open to creating different options for students within schools and multiple ways for parents, educators, and community members to connect with each other.

Today, school districts, schools, teachers, parents, and communities are exploring many different ways to educate children, including charter schools, vouchers, and home schooling. Although the effectiveness of these approaches is not yet clear, we may learn a great deal from them. On the other hand, we must ensure that, in the process of increasing the number of educational options available for parents to choose from, we do not jeopardize resources designated for public education. We must preserve the public school as a "public space" for negotiating differences.

When a community agrees to find ways to educate all children well, it is unified by the common purpose. If concerned community members had known how to do this, they would have done so long ago. Now more than ever, we need to unleash the creativity and imaginations of everyone who cares about the welfare of children. Many of the problems we face today are the same ones we have tried and tried again to solve without success. Instead of clinging stubbornly to familiar structures and traditional ways, we need to explore the unknown and undiscovered.

LAST THOUGHTS: THE DEMOCRATIC IDEAL

In a time of exploding change—with personal lives being torn apart, the existing social order crumbling, and a fantastic new way of life emerging on the horizon—asking the very largest of questions about our future is not merely a matter of intellectual curiosity. It is a matter of survival.

—Alvin Toffler, *The Third Wave*

We pride ourselves on living in a democratic society, and, after September 11, 2001, the democratic ideal became more treasured than we would have thought possible on September 10. Those who were born after World War II have never seen such patriotic fervor so openly displayed. For a time, racial, ethnic, political, religious, and other differences seem less important than national unity. These feelings, however, will not last.

Despite the many positive aspects of a nation unified as the United States has been since terrorists attacked the World Trade Towers and the Pentagon, this unity has blinded us to the many ways that our democracy serves some citizens and not others. TV pictures show us children starving in Afghanistan but not the poor children without enough food or safe homes in our own communities. "At the very least," Jane Roland Martin says, "children need to love and be loved. They need to feel safe and secure and at ease with themselves and others. They need to experience intimacy and affection. They need to be perceived as unique individuals and treated as such."[33] Helping children feel emotionally safe and secure is clearly more difficult today than it was on September 10, 2001, but we know how to make sure every child has food, clothing, and shelter. Satisfying the most basic needs of its youngest members should be a high priority in a democratic society.

If we want to practice rather than only preach the democratic ideal, then we need to ask: What do *all of our* children need to have promising futures? Even those people who are not concerned with ideals should realize that the education all children get today—not just their own children—has practical consequences for everyone. If today's children become adults who cannot hold jobs, who rob others to support drug habits, who abuse or desert their children, who sleep on city streets or in prison cells, the lives of everyone will be affected. In one way or another, now or later, we will pay. In what kind of future society do we want to live?

The future will be in the hands of the children who are in school today. What are we doing to prepare them for the challenges they will face? What can we do? What should we do? These are questions people in every community must ask and answer in their own ways.

In an ideal world, all children would have warm homes with parents who care for them, feed them, clothe them, spend time with them, make them feel special, and love them. They would attend schools that educated them to meet all of life's challenges. They would live in communities where the needs of children came first. In a democratic society, John Dewey said, "What the best and wisest parent wants for his own child, that must be what the community wants for all its children."[34] Whose life would not be improved if the community becomes a home for all of our children?

Resources

WWW.ED.GOV

U.S. Department of Education
400 Maryland Avenue, SW
Washington, DC 20202–0498
800-USA-LEARN

A wealth of information is available through this website. If you'd like to get e-mail information from the Department of Education, you can subscribe to (or unsubscribe from) EDInfo by sending an e-mail message to: listproc@inet.ed.gov. Then write either SUBSCRIBE EDINFO YOURFIRSTNAME YOURLASTNAME in the message or write UNSUBSCRIBE EDINFO. (If you have a signature block, please turn it off.) Then send it!

OTHER HELPFUL GOVERNMENT SITES INCLUDE:

www.ed.gov/free: This collaboration of more than 35 federal agencies makes hundreds of Internet-based education resources easier for student and teachers to access. The U.S. Secretary of Education said it "offers one-stop shopping for a treasure trove of historical documents, scientific experiments, mathematical challenges, famous paintings."

www.ed.gov/Programs/EROD: The Education Resource Organization Directory helps you to identify and contact organizations providing information and assistance on a broad range of education related topics.

www.eric.ed.gov: ERIC: Educational Resources Information Center, a national information system designed to provide ready access to an extensive body of education-related literature.

www.ed.gov/offices/OIIA/Hispanic: White House Initiative of Educational Excellence for Hispanic Americans.

www.ed.gov/offices/OIIA/spanishresources: A variety of free publications are offered through the Department of Education for Spanish speakers and individuals or communities who serve Hispanics.

www.npin.org: National Parent Information Network (NPIN) is a project of the ERIC system whose mission is to provide access to research-based information about the process of parenting and about family involvement in education.

pfie.ed.gov: The Partnership for Family Involvement in Education was established by the U. S. Department of Education. Business, school, community and religious groups commit to increasing family participation in children's learning.

HELPFUL ORGANIZATIONS AND WEBSITES

While many of these organizations are not specifically devoted to school, family, and community partnerships, they have related resources. Many of these organizations publish newsletters, magazines, or journals and offer books for purchase. Try the websites first.

Educators' Associations

American Association of School Administrators
1801 North Moore Street
Arlington, VA 22209–1813
703–528–0700
888–782–2272 (to order publications)
www.aasa.org

American Council on the Teaching of Foreign Languages
Six Executive Plaza
Yonkers, NY 10701
914–936–8830
www.actfl.org

American Educational Research Association (AERA)
1230 17th Street, NW
Washington, DC 20036
www.aera.net

Family, School, Community Special Interest Group of AERA
www.csoes.jhu.edu/fscp/default.htm

American Federation of Teachers (AFT)
555 New Jersey Avenue, NW
Washington, DC 20001
202–393–5676
www.aft.org

Association for Supervision and Curriculum Development (ASCD)
1703 North Beauregard Street
Alexandria, VA 22311–1714
703–578–9600

800-933-ASCD
www.ascd.org

Council for Exceptional Children (Special Education)
111 North Glebe Road, Suite 300
Arlington, VA 22201-5704
888-CEC-SPED
703-620-3660
www.cec.sped.org

Council of State School Officers
1 Massachusetts Avenue, NW, Suite 700
Washington, DC 20001-1431
202-408-5505
www.ccsso.org

Education Commission of the States (ECS)
700 Broadway, Suite 1200
Denver, CO 80202-3460
303-299-3600
www.ecs.org

Music Teachers National Association
617 Vine Street, Suite 1432
Cincinnati, OH 45202
www.mtna.org

National Alliance of Black School Educators
310 Pennsylvania Avenue, SE
Washington, DC 20003
202 608 6319
www.nabse.org

National Art Education Association
1916 Association Drive
Reston, VA 22091-1590
703-860-8000
www.naea-reston.org

National Association of Secondary School Principals (NASSP)
1904 Association Drive
Reston, VA 22091-1537
703-860-0200
www.nassp.org

National Council for the Social Studies
8555 Sixteenth Street, Suite 500
Silver Springs, MD 20910

301-588-1800
www.ncss.org

National Council of Teachers of English
111 W. Kenyon Road
Urbana, IL 61801-1096
800-369-6283
www.ncte.org

National Council of Teachers of Mathematics
1906 Association Drive
Reston, VA 22091-9988
703-620-9840
www.nctm.org

National Education Association
1201 16th Street, NW
Washington, DC 20036
202-833-4000
www.nea.org

National School Boards Association
1680 Duke Street
Alexandria, VA 22314-9900
703-838-6722
www.nsba.org

National Science Teachers Association
1840 Wilson Boulevard
Arlington, VA 22201-3000
703-243-7100
www.nsta.org

Phi Delta Kappa
408 N. Union Street
P.O. Box 789
Bloomington, IN 47402-0789
1-800-766-1156
812-339-1156
www.pdkintl.org

School Reform

Achieve
Washington office:
400 North Capitol Street, Suite 351
Washington, DC 20001
202-624-1460
www.achieve.org

Massachusetts office:
8 Story Street, 1st fl.
Cambridge, MA 02138
617-496-6300

Annenberg Institute for School Reform
Brown University, Box 1985
Providence, RI 02912
401–863–7990
www.annenberginstitute.org

The Boyer Center
Messiah College
One College Avenue, Box 3024
Grantham, PA 17027
717–796–5077
www.boyercenter.org

Business Coalition for Education Reform
c/o National Alliance of Business
1201 New York Avenue, NW, Suite 700
Washington, DC 20005–3917
800–787–2848
www.bcer.org

The Business Roundtable
1615 L Street, NW, Suite 1100
Washington, DC 20036
202–872–1260
www.brtable.org

Center for School Change
Hubert H. Humphrey Institute of Public Affiars
University of Minnesota, Twin Cities
301 19th Avenue South, Room 234
Minneapolis, MN 55455
612–626–1834
www.hhh.umn.edu/centers/school-change

Coalition of Essential Schools
1814 Franklin Street, Suite 700
Oakland, CA 94612
510–433–1451
www.essentialschools.org

Comer School Development Program
55 College Street
New Haven, CT 06510
203–737–1020
info.med.yale.edu/comer

Council for Basic Education
1319 F Street, NW, Suite 900
Washington, DC 20004–1152

202-347-4171
www.c-b-e.org

National Education Goals Panel
1255 22nd Street, NW, Suite 502
Washington, DC 20037
202-724-0015
202-632-0957
www.negp.gov

National Governors' Association
444 North Capitol Street
Washington, DC 20001-1512
202-624-5300
www.nga.org

NECA/Teaching for Change
P.O. Box 73038
Washington, DC 20056
800-763-9131
www.teachingforchange.org

North Central Regional Educational Laboratory
1120 East Diehl Road, Suite 200
Naperville, Illinois 60563
www.ncrel.org
www.ncrel.org/sdrs/pidata/pi0over.htm (Parent Involvement Data Base)

Family Involvement

Academic Development Institute
The Center for the School Community
Family Study Institute
Illinois Family Education Center
121 N. Kickapoo Street
Lincoln, IL 62656
www.adi.org

The Center for Family Involvement in Schools
Rutgers, the State University of New Jersey
732-445-1287
www.rci.rutgers.edu/~cfis

Center on School, Family, and Community Partnerships
Johns Hopkins University
3003 North Charles Street, Suite 200
Baltimore, MD 21218
410-516-8800
www.csos.jhu.edu

Family Friendly Schools
13080 Brookmead Drive
Manassas, Virginia 20112
800-648-6082
www.familyfriendlyschools.org

Family Involvement Network of Educators (FINE)
Harvard Family Research Project
Harvard Graduate School of Education
Longfellow Hall, Appian Way
Cambridge, MA 02138
617-495-9108
www.finenetwork.org

The Home and School Institute
Mega Skills Education Center
1500 Massachusetts Avenue, NW
Washington, DC 20005
202-466-3633
www.megaskillshsi.org

Institute for Responsive Education
21 Lake Hall
Northeastern University
Boston, MA 02115
617 373 2595
www.resp-ed.org

The National Coalition for Parent Involvement in Education
3929 Old Lee Highway, Suite 91-A
Fairfax, VA 22030-2401
703-359-8973
www.ncpie.org

The National PTA
Chicago office:
330 North Wabash Avenue, Suite 2100
Chicago, IL 60611-3690
312-670-6782
www.pta.org (e-mail: info@pta.org)

Washington office:
1090 Vermont Avenue, NW, Suite 1200
Washington, DC 20005-6790
202-289-6790

Parents as Teachers National Center
10176 Corporate Square Drive, Suite 230
St. Louis, MO 63132
314-432-4330
www.patnc.org

Parents for Public Schools
1520 North State Street
Jackson, MS 39202

601-354-1220
www.parents4publicschools.com

Project Parents
46 Beach Street, Suite 502
Boston, MA 02111
617-451-0360
www.projectparents.org

Advocacy Groups

ACORN—Association of Community Organizations for Reform Now
ACORN National
88 3rd Avenue
Brooklyn, NY 11217
877-55ACORN
www.ACORN.org

2101 South Main Street
Little Rock, AR 72206
501-376-3952

1018 West Roosevelt Street
Phoenix, AZ 85007
602-254-5299

1024 Elysian Fields Avenue
New Orleans, LA 70117
504-943-0044

1453 Dorchester Avenue
Boston, MA 02122
617-436-7100

739 8th Street South East
Washington, DC 20003
202-547-2500

Center for Law and Education (CLE)
1875 Connecticut Avenue, NW Suite 510
Washington, DC 20009
202-986-3000
www.cleweb.org

The Children's Defense Fund
25 E Street, NW
Washington, DC 20001
202-628-8787
www.childrensdefense.org

Mexican American Legal Defense and Educational Fund
Community Education and Public Policy
634 South Spring Street
Los Angeles, CA 90014
213-629-0839
www.maldef.org

National Black Child Development Institute
1101 15th Street, NW, Suite 900
Washington, DC 20005
202-833-2220
www.nbcdi.org

National Center for Fair and Open Testing (FairTest)
342 Broadway
Cambridge, MA 02139
617-864-4810
www.fairtest.org
www.nclr.org

National Coalition of Education Activists
P.O. Box 679
Rhinebeck, NY 12572
845-876-4580
http://members.aol.com/nceaweb

National Council of Jewish Women
Center for the Child
53 W. 23rd Street, 6th Floor
New York, NY 10010
212-645-4048
www.ncjw.org

National Council of La Raza
1111 19th Street, NW, Suite 1000
Washington, DC 20036
202-785-1670
www.nclr.org

National Urban League
120 Wall Street
New York, NY 10005
212-558-5300
www.nul.org

PACER Center, Inc: Parent Advocacy Coalition for Educational Rights for Parents of Children with Disabilities
4826 Chicago Avenue South
Minneapolis, MN 55417
1-888-248-0822
pacer@pacer.org

Parents United for Responsible Education (PURE)
307 Dearborn Street, #515

Chicago, IL 60605
312–461–1994
www.pureparents.org

Quality Education for Minorities (QEM) Network
1818 N Street, NW
Washington, DC 20036
202–659–1818
www.qemnetwork.qem.org

Community Engagement

America's Promise—The Alliance for Youth
909 North Washington Street, Suite 400
Alexandria, VA 22314
888–559–6884
www.americaspromise.org

Civic Practices Network (CPN)
Center for Human Resources
Heller School for Advanced Studies in Social Welfare
Brandeis University
60 Turner Street
Waltham, MA 02154
617–736–4890
www.cpn.org

Communities in Schools, Inc.
277 South Washington Street
Alexandria, VA 22314
703–518–2557
www.cisnet.org

National Center for Community Education
1017 Avon Street
Flint, MI 48503
810 238 0463
www.nccenet.org

National Civic League
Washington office:
1319 F Street, NW
Washington, DC 20004
800–308–9414
202–783–2961
www.ncl.org

Denver office:
1445 Market Street, Suite 300
Denver, CO 80202
800–223–6004
303–571–4343

National Community Education Association
3929 Old Lee Highway, #91-A

Fairfax, VA 22042
703–359–8973
www.ncea.com

Public Agenda
6 East 39th Street
New York, NY 10016
212–686–6610
www.publicagenda.org

Public Education Network
601 13th Street, NW, Suite 900 North
Washington, DC 20005
202–628–1893
www.publiceducation.org

Notes

CHAPTER 1

1. National Commission on Excellence in Education, *A Nation at Risk* (Washington, DC: U.S. Department of Education, 1983), p. 5.
2. Michael G. Fullan, *The Meaning of Educational Change* (New York: Teachers College Press, 1982), p. 245.
3. Lawrence A. Cremin, *Public Education* (New York: Basic Books, 1976).
4. Robert Friedman and David Meltzer, "Family Factors in Learning Disability," in *Family Roots of School Learning and Behavior Disorders*, ed. Robert Friedman (Springfield, IL: Charles C Thomas, 1973), pp. 55–67; Sanford M. Dornbusch and Philip L. Ritter, "When Effort in Schools Does Not Produce Better Grades: A Family Environment Affects School Practice," in *Parent-Adolescent Relationships*, ed. Brian K. Barber and Boyd C. Rollins (New York: University Press of America, 1990), pp. 75–93.
5. Robert Friedman, "School Behavior Disorder and the Family," in *Family Roots of School Learning and Behavior Disorders*, pp. 68–69.
6. Saul Brown, "Family Experience and Change," in *Family Roots of School Learning and Behavior Disorders*, p. 36.
7. *Strong Families, Strong Schools: Building Community Partnerships for Learning* (Washington, DC: U.S. Department of Education, 1994), p. 2.
8. Carol Ames et al., *Parent Involvement: The Relationship between School-to-Home Communication and Parents' Perceptions and Beliefs*, ERIC document #ED362271, Report No. 15, 1993.
9. Anne Wescott Dodd and Jean L. Konzal, *Making Our High Schools Better: How Parents and Teachers Can Work Together* (New York: St. Martin's Press, 1999).
10. Jack Edward Wilhite, "The Response of a Small Rural Community in the Midwest to the Recommendations of the Anglo-American Conference on the Teaching of English" (Ph.D. dissertation: University of Illinois at Urbana-Champaign, 1973), p. 224.
11. Dodd and Konzal, *Making Our High Schools Better*.
12. "Number of Home-Schoolers Shows Steep Rise in Maine," *Portland Press Herald*, August 28, 2001, p. 2B.
13. Catherine Tsai, "More Grandparents Getting Second Chance at Raising Babies," Associated Press story, *The Times Record* (Brunswick, ME) (July 7, 2001): 13.
14. National Center for Education Statistics, *The Condition of Education 2001* (Washington, DC: U.S. Department of Education, 2001), 93. Online at http://nces.ed.gov/pubsearch—key in publication number 2001072.

15. Andy Hargreaves, "Beyond Anxiety and Nostalgia: Building a Social Movement for Educational Change," *Phi Delta Kappan* 82, no. 5 (September 2001): 377.

CHAPTER 2

1. Jean L. Konzal, *Our Changing Town, Our Changing School: Is Common Ground Possible?*, unpublished doctoral dissertation, University of Pittsburgh, 1995, 103.
2. Thomas F. Green, *The Activities of Teaching* (New York: McGraw Hill, 1971), 55.
3. M. Frank Pajares, "Teachers' Beliefs and Educational Research: Cleaning Up a Messy Construct," *Review of Educational Research* 62:3 (1992): 321.
4. Milton Rokeach, *The Open and Closed Mind* (New York: Basic Books, 1960); *Beliefs, Attitudes and Values* (San Francisco: Jossey-Bass, 1968).
5. Paul Ernest, "The Knowledge, Beliefs and Attitudes of the Mathematics Teacher: A Model," *Journal of Education for Teaching* 15:1 (1989): 30.
6. Milton Rokeach, *The Open and Closed Mind*, 38.
7. Anne Wescott Dodd, *Parents as Partners in Learning: Their Beliefs about Effective Practices for Teaching and Learning High School English*, unpublished doctoral dissertation, University of Maine, 1994.
8. Konzal, 101.
9. William W. Cutler III, *Parents and Schools: The 150-Year Struggle for Control in American Education* (Chicago: University of Chicago Press, 2000), p. 3.
10. Paul Houston, "Superintendents for the 21st Century," *Phi Delta Kappan* 82:6 (February 2001): 431.
11. Lawrence A. Cremin, *Public Education* (New York: Basic Books, 1976).
12. Shirley Brice Heath, *Ways with Words* (New York: Cambridge University Press, 1983).
13. Michael Young and Patrick McGeeney, *Learning Begins at Home: A Study of a Junior School and Its Parents* (London: Routledge & Kegan Paul, 1968), 102.
14. Daphne Johnson and Elizabeth Ransom, *Family and School* (London: Croom Helm, 1983), 89.
15. Margaret A. Gibson, *Accommodation without Assimilation: Sikh Immigrants in an American High School* (Ithaca, NY: Cornell University Press, 1988).
16. William Kay, *Moral Education: A Sociological Study of the Influence of Society, Home and School* (London: Linnet Books, 1975).
17. Sara Lawrence Lightfoot, *Worlds Apart: Relationships Between Families and Schools* (New York: Basic Books, 1978); Daphne Johnson and Elizabeth Ransom, *Family and School*; Jean Anyon, "Social Class and the Hidden Curriculum of Work," *Journal of Education* 162 (1980): 67–92, in *Society and Education*, 7th ed., Daniel U. Levine and Robert J. Havighurst (Boston, MA: Allyn and Bacon, 1989).
18. Anne Wescott Dodd and Jean L. Konzal, *Making Our High Schools Better: How Parents and Educators Can Work Together* (New York: St Martin's Press, 1999).
19. Konzal; also see chapter 4 in this book.
20. Gibson.
21. Lightfoot, 165.
22. Young and McGeeney, 74.

23. Johnson and Ransom, 8.
24. Konzal, 100–101.
25. Cutler, 5.
26. Dodd and Konzal.
27. Harriet Tyson, "A Load off the Teachers' Backs: Coordinated School Health Programs," *Phi Delta Kappan* 80:5 (January 1999): K3.
28. Michelle Fine, "[Ap]parent Involvement: Reflections on Parents, Power, and Urban Public Schools," *Teachers College Record* 94:4 (1993): 682.
29. Fine, 682.

CHAPTER 3

1. The information for this chapter came from personal interviews, telephone calls, and written correspondence with parents and school officials as well as from local newspaper articles. The newspapers are not named to protect the identity of those involved. All names, including the name of the community, are pseudonyms.

CHAPTER 4

1. The names of the school, towns, and people involved are pseudonyms. Some information in this chapter comes from the school's website and articles in the newspaper that routinely covers events in the three towns in which SRHS students and their parents live. Other information came from e-mail correspondence, personal interviews, and observations at the school during the 2000–2001 school year. Credit also must be given for some material quoted from an unpublished manuscript by Jesse Reich completed for a college course requirement in 2001. In order to preserve the anonymity of the school and the people involved, specific sources are not named.

CHAPTER 5

1. Thomas P. Carey, letter sent to all superintendents in the state of Pennsylvania, October 14, 2001.
2. Jane Roland Martin, *The Schoolhome* (Cambridge, MA: Harvard University Press, 1992), pp. 170–171.
3. See ibid. for a complete explanation of Martin's philosophy and the basis for it.
4. Theodore R. Sizer and Nancy Foust Sizer, *The Students Are Watching: Schools and the Moral Contract* (Boston: Beacon Press, 1999), p. 121.
5. Nan Henderson, "Connectedness Is Crucial," *Student Assistance Journal* (January/February 1999): 30.
6. Paul Houston, "Superintendents for the 21st Century," *Phi Delta Kappan* 82, no. 6 (February 2001): 431.
7. Sam Redding, "The Community of the School," in *The Community of the School*, ed. S. Redding and Lori G. Thomas (Kickapoo, IL: Academic Development Institute, 2001), p. 6.
8. Robert Bellah et al., *Habits of the Heart: Individualism and Commitment in American Life* (New York: Harper Collins, 1985); Amitai Etzioni, *The Spirit of Community* (New York: Crown, 1993).

9. Mary Henry, *Parent-School Collaboration* (Albany: State University of New York Press, 1995), p. 17.
10. James Gleick, *Chaos: Making a New Science* (New York: Penguin, 1987), p. 8.
11. Alvin Toffler, *The Third Wave* (New York: Morrow, 1980), p. 119.
12. Ervin Tazlo quoted in ibid., p. 287.
13. Ira J. Gordon, "The Effects of Parent Involvement in Schooling," in *Parents and Schools*, ed. R. S. Brandt (Alexandria, VA: Association for Supervision and Curriculum Development, 1979), reprinted in *Parent Involvement in Education*, Iowa State Department of Education on the Web at npin.org/library/pre1998/n00321/n00321.html.
14. Joyce L. Epstein, "School/Family/Community Partnerships: Caring for the Children We Share," *Phi Delta Kappan* 76, no. 9 (1995): 702.
15. Houston, "Superintendents for the 21st Century," p. 431.
16. Erwin Flaxman, "The Promise of Urban Schooling," *Eric Review* 8, no. 2 (Winter 2001): 9.
17. Ibid.
18. D. Mathews, *Is There a Public for Public Schools?* (Dayton, OH: Charles F. Kettering Foundation, 1996), p. 25, quoted in Timothy Collins and Erwin Flaxman, "Improving Urban and Rural Schools and Their Communities," *Eric Review*, 8, no. 2 (Winter 2001): 3.
19. William W. Cutler III, *Parents and Schools: The 150-Year Struggle for Control in American Education* (Chicago: University of Chicago Press, 2000), p. 13.
20. Lisa Delpit, *Other People's Children: Cultural Conflict in the Classroom* (New York: New Press, 1996).
21. William Ayers, *To Teach: The Journey of a Teacher* (New York: Teachers College Press, 1993), pp. 64–65.
22. Emily Style, "Curriculum as Window and Mirror," in *Listening for All Voices: Gender Balancing the School Curriculum*, proceedings of a conference held at Oak Knoll School, Summit, NJ, 1988, pp. 6–12.
23. Janet Atkin and John Bastiani (with Jackie Goode), *Listening to Parents: An Approach to the Improvement of Home/School Relations* (London: Croom Helm, 1988), 18.
24. Anne Wescott Dodd and Jean L. Konzal, *Making Our High Schools Better: How Parents and Educators Can Work Together* (New York: St. Martin's Press, 1999).
25. Andy Hargreaves, "Beyond Anxiety and Nostalgia: Building a Social Movement for Educational Change," *Phi Delta Kappan* 82, no. 5 (January 2001): 377.
26. Michele Foster, *Black Teachers on Teaching* (New York: New Press, 1997), p. 146.
27. National Council of Teachers of English, "Volunteering Can Be Good for You," *Council-Grams: News and Information for Leaders of NCTE*, 64, no. 1 (September 2001), e-mail version, p. 1.
28. "Value of Volunteer Time on the Rise," *Council-Grams: News and Information for Leaders of NCTE*.
29. Flaxman, "The Promise of Urban Schooling," pp. 12–13.
30. John O'Neill, "On Schools as Learning Organizations: A Conversation with Peter Senge," *Educational Leadership* 52, no. 7 (April 1995): 22.
31. Michelle Fine, "[Ap]parent Involvement: Reflections on Parents, Power, and Urban Public Schools," *Teachers College Record* 94, no. 4 (1993): 20; online at www.tcrecord.org.

32. John Dewey, *Democracy and Education* (New York: Macmillan, 1916/1944), p. 9.
33. Meredith Maran, "Damage Control," *Teacher Magazine* (August/September 2001): 25–37.
34. Thomas J. Sergiovanni, *Building Community in Schools* (San Francisco: Jossey Bass, 1994), p. 4.
35. J. W. Gardner, "School and Community," *Community Education Journal* 23, nos. 1–2 (1995/1996): 6–8, quoted in Collins and Flaxman, "Improving Urban and Rural Schools and Their Communities," p. 4.
36. National Center for Education Statistics, *The Condition of Education 2001* (Washington, DC: U.S. Department of Education, 2001), p. 115. Online at nces.ed.gov/pubsearch—key in publication number 2001072.
37. Bruce Joyce and Emily Calhoun, "School Renewal: An Inquiry, Not a Formula," *Educational Leadership* 52, no. 7 (April 1995); online at www.ascd.org.
38. "What Should Parents Know about Schools as Community Learning Center?" online at www.ed.gov.
39. B. A. Miller, *The Role of Rural Schools in Rural Community Development* (Charleston, WV: ERIC Digest, ERIC document ED 384 479), in Timothy Collins, "Rural Schools and Communities: Perspectives on Interdependence," *Eric Review* 8, no. 2 (Winter 2001): 18.
40. Collins, "Rural Schools and Communities."
41. Mary Ann Zehr, "One School, Two Cultures," *Education Week*, 21, no. 2 (September 12, 2001), pp. 35–39.
42. National Center for Community Education, "Parents Empowered to Take the Lead in Pharr, Texas," *The Community Educator* 3, no. 3 (Fall 2001): 4. According to a note in the article, the excerpt quoted was taken from the Association for Supervision and Curriculum Development publication, "Beyond the Classroom," by Yolanda Castillo. No other information was provided.
43. Ibid.
44. Edgar Allen Beem, "We'll Do It Our Way," *Down East* 48, no. 4 (November 2001): 69.
45. "What Should Parents Know about Schools as Community Learning Center?"

CHAPTER 6

1. The annual survey by Phi Delta Kappa International, an educators' association, has consistently shown that parents rate their children's school higher than other schools.
2. For example, Wayne K. Hoy has examined trust among faculty members as a variable in effective school organizations since the mid-1980s and has mentored many of his doctoral students in their studies of trust. Most recently, this research agenda has begun to look at the issue of teacher trust of parents and students. See Roger D. Goddard, Megan Tschannen-Moran, and Wayne K. Hoy, "A Multilevel Examination of the Distribution and Effects of Teacher Trust in Students and Parents in Urban Elementary Schools," *The Elementary School Journal* 102, no. 1 (2001): 3–17. Also see Kimberly Sue Adams, "Trust Within the Home-School Relationship: An Empirical Investigation of Parent and Teacher Perspectives." Ph.D. dissertation, University of Minnesota, 1997; and Michelle D. Young, "Importance of Trust in Increasing Parental Involve-

ment and Student Achievement in Mexican American Communities," paper presented at the Annual Meeting of the American Association of Educational Researchers, San Diego, CA, 1998.
3. For example: C. J. Dunst, C. Johanson, T. Rounds, C. M. Trivette, and D. Hamby, "Characteristics of Parent and Professional Partnerships, in *Home-School Collaboration: Enhancing Children's Academic and Social Competence*, ed. S. L. Christenson and J. C. Conoley (Colesville, MD: NASP, 1992), pp. 157–174, cited in Adams, "Trust within the Home-School Relationship," p. 23. Sharon M. Allen, Ray H. Thompson, Michael Hoadley, and Jeri Engelking, "What Teachers Want from Parents and What Parents Want from Teachers: Similarities and Differences," paper presented at the Annual Meeting of the American Educational Research Association, Chicago, IL, 1997.
4. A. K. Mishra, "Organizational Responses to Crisis: The Centrality of Trust," in *Trust in Organizations*, ed. R. Kramer, and T. Tyler (Thousand Oaks, CA: Sage, 1996), cited in Megan Tschannen-Moran and Wayne Hoy, "Faculty Trust: A Conceptual and Empirical Analysis," paper presented at the Annual Meeting of the American Educational Research Association, Chicago, IL, 1997, p. 2.
5. John K. Rempel, John G. Holmes, and M. P. Zanna, "Trust in Close Relationships," *Journal of Personality and Social Psychology* 49, no. 1 (1985): 93–112, cited in Tschannen-Moran and Hoy, "Faculty Trust," pp. 24–25.
6. Kathy Murfitt, "Addressing the State of Parent-Teacher Relations." *Washington Post*, May 22, 2001, CO4. See www.washingtonpost.com/wp-dyn/education/teachers.
7. Adams, "Trust within the Home-School Relationship," p. 25.
8. Quotes come from discussions with the participants of the Greenbrook Elementary School Kindergarten Parent Journal Writing project following the tragic events in Columbine.
9. Jean L. Konzal, "Our Changing Town, Our Changing School: Is Common Ground Possible," Ph.D. dissertation, University of Pittsburgh, 1995, p. 87.
10. Linda Perlstein, "Suspicious Minds" (Washington D.C.: Washington Post July 22, 2001), p. 1.
11. Murfitt, "Addressing the State of Parent-Teacher Relations," p. 2.
12. From transcript of interview conducted by Konzal for "Our Changing Town, Our Changing School."
13. Murfitt, "Addressing the State of Parent-Teacher Relations," p. 2.
14. Colleen Pohlig, "Schools Spelling Out Need for Parental Civility" (Seattle, WA: Seattle Times Company), p. 2
15. Ibid., p. 110.
16. Goddard, Tschannen-Moran, and Hoy, "A Multilevel Examination of the Distribution and Effects of Teacher Trust," p. 15.
17. Ibid.
18. Coauthor Jean Konzal has her undergraduate students interview teachers. One of the questions is: What are the biggest problems you face as a teacher? This quote is representative of quotes collected during these interviews.
19. Ibid.
20. Peter Senge, *The Fifth Discipline: The Art and Practice of the Learning Organization* (New York: Doubleday, 1990).

21. William Boyd, "The Public, the Professionals, and Educational Policy-Making, Who Governs?" *Teachers College Record* 77 (May 1976): 539–577.
22. Jeannie Oakes, *Keeping Track: How Schools Structure Inequality* (New Haven, CT: Yale University Press, 1985).
23. Tim Zukas, "An Open Letter from a Parent: So You Want to Implement Reform" *Educational Leadership* (April 2000): 54.
24. Editorial, "Classroom 'Socialism,' or a Lesson in Sharing?" *Bucks County Courier Times*, September 15, 2000, p. 14A.
25. Anne Wescott Dodd and Jean Konzal, *Making Our High Schools Better: How Parents and Teachers Can Work Together* (New York: St. Martin's Press, 1999).
26. For instance, Wayne K. Hoy, C. J. Tarter, and L. Witkoskie, "Faculty Trust in Colleagues: Linking the Principal with School Effectiveness," *Journal of Research and Development In Education* 26, no. 1 (1992): 38–45.
27. David Bensman, *Building School-Family Partnerships in a South Bronx Classroom* (New York: National Center for Restructuring Education, Schools and Teaching, Teachers College, Columbia University, 1999).
28. Ibid.
29. Michelle Galley, "Chicago to Size Up Parents with 'Checklists,'" *Education Week*, May 31 (2000): 3.
30. Ibid.
31. Joyce Epstein, *School, Family, and Community Partnerships: Preparing Educators and Improving Schools* (Boulder, CO: Westview Press, 2001).
32. Bensman, *Building School-Family Partnerships*, p. iii.
33. Roland S. Barth, *Improving Schools from Within: Teachers, Parents, and Principals Can Make the Difference* (San Francisco: Jossey-Bass, 1990), p. 122.
34. For instance, Goddard, Tschannen-Moran, and Hoy, "A Multilevel Examination of the Distribution and Effects of Teacher Trust."

CHAPTER 7

1. Greenbrook Kindergarten Parent Readers' Theater Script, 2000–2001.
2. Concurrent with this site council summer planning meeting, the school, along with other schools in the area, joined forces with The College of New Jersey to create a Professional Development School Network—a collaborative effort to reform the way schools of education and K–12 schools work together in the preparation of new teachers and the continued professional development of current teachers. A new faculty member at the college at the time, coauthor Jean Konzal, who had an interest in parent/school relationships and the construction and use of readers' theater scripts, was invited to collaborate on this effort. In addition to consulting to the Kindergarten Journal Writing project, Konzal also has supervised student teachers placed at the school for the past four years.
3. Parents who participated in the Kindergarten Journal Writing project were invited to write short essays about their experiences as parents in the school. Three parents, Carol Desmond, Miriam Lally, and Shari Rothstein, accepted this invitation. Quotes from their essays are woven throughout this school portrait.
4. Mary Henry, *Parent-School Collaboration: Feminist Organizational Structures and School Leadership* (Albany: State University of New York Press, 1996).

5. The events of September 11 has brought some changes to Greenbrook. Because of security concerns, a large sign now appears on the front entrance clearly stating that visitors must report to the office and that trespassers will be prosecuted.
6. Ernest Boyer, *High School: A Report on Secondary Education in America* (New York: Harper & Row, 1983).
7. Greenbrook Kindergarten Parent Readers' Theater Script, 1998–1999.

CHAPTER 8

1. See groups.yahoo.com/group/crossroadssc/messages, January 14, 2000, message 362.
2. Ibid., message 363.
3. See groups.yahoo.com/group/crossroadssc/messages, January 15, 2000, message 371.
4. Ibid., message 365.
5. Ibid., message 365.
6. Ibid., message 371.
7. The Coalition of Essential Schools, based on the work of Ted Sizer and his associates, is a nation-wide network of schools committed to school reform based on ten essential principles. Go to www.essentialschools.org for more information.
8. Crossroads School, *A Handbook for Crossroads School*, 2000, p. 4.
9. Emily Style, "Curriculum as Window and Mirror," in *Listening for All Voices: Gender Balancing the School Curriculum*, proceedings of a conference held at Oak Knoll School, Summit, NJ, 1988, pp. 6–12.
10. Crossroads School handbook, p. 5.
11. Ana Chanlatte, "Discipline . . . with Love and Caring," in ed. Michelle Fine, *Talking Across Boundaries: Participatory Evaluation Research in an Urban Middle School* (New York: Bruner Foundation, 1996), p. 57.
12. Ibid., pp. 57–61.
13. Crossroads School handbook, 17.
14. Comments by parents not attributed to a Yahoo! Groups message are the result of interviews conducted with a number of parents of present and past Crossroads students.
15. During the 2001–2002 school year due to pressure from the district office, the school had to admit more children than it had planned, resulting in larger advisory groups.
16. Jianzhong Xu, *Reaching Out to Other People's Children in an Urban Middle School: The Families' Views* (New York: National Center for Restructuring Education, Schools and Teaching, 2000).
17. Ibid., p. 62.
18. Thomas J. Sergiovanni, "Value-Driven Schools: The Amoeba Theory," in *Organizing for Learning: Toward the 21st Century*, ed. Harry Walberg and John Lane (Reston, VA: National Association of Secondary School Principals, 1989), pp. 31–40.
19. Ann Wiener, "Participatory Leadership: Inside the Circle, Outside the Circle, Expanding the Circle," in *Talking Across Boundaries*, p. 35.
20. Ibid.

21. Michelle Fine, "At a Crossroads," in *Talking Across Boundaries*, p. 13.
22. Mary E. Henry, *Parent-School Collaboration: Feminist Organizational Structures and School Leadership* (Albany: State University of New York Press, 1996).
23. Quoted in Kathleen F. Malu, "A Study of Parent Concerns for Children's Work in Schools," in *Talking Across Boundaries*, p. 65.
24. Ibid., p. 67.
25. Jianzhong Xu, *Reaching Out to Other People's Children*, p. 47.

CHAPTER 9

1. The notes cite the sources for the practices and programs in this chapter. Although individual quotations do not have separate citations, any words appearing in quotation marks were taken directly from the source indicated for that section.
2. For more information about the Boyer Center and about these exemplary schools, contact: The Boyer Center, Messiah College, P.O. Box 3024, One College Avenue, Grantham, PA 17027. Telephone: (717) 796-5077. E-mail: boyerctr@messiah.edu. Website: www.boyercenter.org.
3. Ernest L. Boyer, *The Basic School: A Community for Learning* (San Francisco: Jossey Bass, 1995).
4. John H. Hollifield, ed., "Parents' Perceptions and Beliefs about School-to-Home Communications Affect Their Involvement in Their Children's Learning," *Research and Development Report* (Baltimore: Center on Families, Communities, Schools, & Children's Learning, Johns Hopkins University), p. 5.
5. Nicole Nichols-Solomon, "Conquering the Fear of Flying," *Phi Delta Kappan* (September 2000): 19–21.
6. Robert C. Johnston, "Polling Parents," *Education Week* 20, no. 37 (May 23, 2001): 5.
7. Nancy S. Grasmuck, "How Maryland Communicates Change," *Educational Leadership* 57, no. 7 (April 2000): 44–47.
8. Teresa Jo Clemens-Brower, "Recruiting Parents and the Community," *Educational Leadership* 54, no. 5 (February 1997): 58–60.
9. Dianna Just, "Parent-Teacher Conferences in the Aisles of Wal-Mart," *The Voice* 3, no. 3, (Fall 1998): 3.
10. Partnership for Family Involvement in Education, "Examples of Families and Schools Working Together to Improve Education." From an unpaginated packet of materials developed for families, educators, and community members. For more information, see Partnership for Family Involvement in Education online at pfie.ed.gov.
11. Ibid.
12. Public Education Fund Network, "KIVA: A Strategy For Public Engagement," *Connections*, 3, no. 2 (Winter 1996): 7. For more information, contact the Public Education Fund Network, 601 Thirteenth Street, NW, Suite 290, North, Washington, DC 20005. Telephone: (202) 628-7460.
13. Wayne Jacobsen, "Why Common Ground Thinking Works," *Educational Leadership*, 57, no. 4 (January 2000): 76–80.
14. Nichols-Solomon, "Conquering the Fear of Flying," p. 21.

15. Luis C. Moll, C. Amanti, D. Neff, and N. Gonzalez, "Funds of Knowledge for Teaching: Using a Qualitative Approach to Connect Homes and Classrooms," *Theory into Practice*, 31, no. 2 (November 1992): 132–141.
16. Susan M. Swap, "Reaching Out to Culturally Diverse Families," unpublished paper, Boston: Wheelock College, 1991, pp. 5–7.
17. Edwards, Patricia A., and Heather M. Pleasants, "Uncloseting Home Literacy Environments: Issues Raised through the Telling of Parent Stories," *Early Child Development and Care*, 127–128 (1997): 27–46.
18. Jessica L. Sandham, "Home Visits Lead to Stronger Ties, Altered Perceptions," *Education Week* 49, no. 14 (December 1, 1999): 6.
19. Sudia Paloma McCaleb, *Building Communities of Learners* (Mahwah: Lawrence Erlbaum Associates, 1997).
20. Jane Spielman, "The Family Photography Project," *Family, School, Community Partnerships Newsletter* (Fall 1998): 3–5.
21. Partnership for Family Involvement in Education, "Examples of Families and Schools Working Together."
22. See equals.lhs.berkeley.edu/.
23. National Network of Partnership—2000 Schools, "School Report: School Wide Interactive Homework Project Involves Families in Enfield, Connecticut," *Type 2* (Johns Hopkins University) 2 (Spring 1997): 6.
24. Illinois Family Education Center, "Chicago Public Schools: Partnering with Parents," *Families & Schools: Building Learning Communities for Parents* 3, no. 2 (Fall 2000): 2.
25. Gerardo R. Lopez, Jay D. Scribner, and Kanya Mahitvanichcha, "Redefining Parent Involvement: Lessons from High-Performing Migrant-Impacted Schools," *American Educational Research Journal* 38, no. 2 (Summer 2001): 253–288.
26. Illinois Family Education Center, "Parent Connection Is Integral to School-Family Partnership," *Families and Schools* 1, no. 2 (Fall 1998): 2.
27. For more information about F.A.S.T., contact Family Service America, 11700 W. Lake Park Drive, Milwaukee, WI 53224. Telephone: (414) 359–1040.
28. Illinois Family Education Center, "F.A.S.T. a Success in Oregon—Mt. Morris," *Families & Schools: Building Learning Communities for Parents* 3, no. 2 (Fall 2000): 5.
29. Illinois Family Education Center, "Teaching English as a Second Language," *Families & Schools: Building Learning Communities for Parents* 3, no. 2 (Fall 2000): 6.
30. Partnership for Family Involvement in Education, "Examples of Families and Schools Working Together."
31. John O'Neil, "Building Schools as Communities," *Educational Leadership* 54, no. 7 (May 1997): 6–10.
32. Linda Jacobson, "First Things First," *Education Week* 20, no. 1 (September 6, 2000): 52–56.
33. Information gathered from www.operationbookworm.com/volunteer.htm and personal communication with Ellen Gordon.
34. Clemens-Brower, "Recruiting Parents and the Community," pp. 58–60.
35. Illinois Family Education Center, "Metro East Parent Connection Opens in East St. Louis Area," *Families & Schools: Building Learning Communities for Parents* 3, no. 2 (Fall 2000): 1.

36. Mitch Dorson, "Against the Grain: An Overview of the Triumphant First Year of the U.S. History Class for Parents," presentation at 1999 Coalition of Essential Schools Fall Forum in Atlanta, GA. For more information, contact Dorson at Catalina Foothills High School, 4300 East Sunrise, Tucson, AZ 85718.
37. Susan Connell Biggs, "Writing Workshops: Linking Schools and Families," *English Journal* 90, no. 5 (May 2001): 45–51.
38. Seymour B. Sarason, *Parental Involvement and the Political Principle: Why the Existing Governance Structure of Schools Should Be Abolished* (San Francisco: Jossey-Bass, 1995), p. 39.
39. Mavis G. Sanders, "Building Family Partnerships That Last," *Educational Leadership* 54, no. 3 (November 1996): 61–66.
40. John H. Hollifield, ed., "Parent-Teacher Action Research: The Larger Meaning," *Research and Development Report* (Baltimore: Center on Families, Communities, Schools, & Children's Learning, Johns Hopkins University) 6 (September 1995): 1–4.
41. Illinois Family Education Center, "School Community Councils Help Grow Strong Minds," (*Families & Schools: Building Learning Communities for Parents* 3, no. 2 (Fall 2000): 10–11.
42. Jacqueline Edmonson, Gregory Thorson, and David Fluegel, "Big School Change in a Small Town," *Educational Leadership* 57, no. 7 (April 2000): 51–53.
43. John Conyers and W. Christine Rauscher, "An Antarctic Adventure," *Educational Leadership* 58, no. 4 (December 2000): 69–72.
44. Summary written by Kit Juniewicz, University of New England, Biddeford, ME.
45. U.S. Department of Education, "Partnership for Family Involvement in Education," *Community Update* 91 (September 2001): 6.
46. Ibid.
47. Partnership for Family Involvement in Education, "East Harlem Tutorial—40 Years of Volunteering and Mentoring Success." See pfie.ed.gov. For more information about this program, contact EHTP, 2050 Second Avenue, New York, NY, 10029. Telephone: (212) 831–0650.
48. Partnership for Family Involvement in Education, "Examples of Families and Schools Working Together."
49. Ibid.
50. Ibid.
51. Betty L. Dixon, "The Great Family Network," *Educational Leadership* 53, no. 7 (April 1996): 27–29.
52. Eva Marx and Daphne Northrop, "Partnerships to Keep Students Healthy," *Educational Leadership* 57, no. 6 (March 2000): 22–24.
53. John H. Hollifield, ed., "Projects at a Glance," *Research and Development Report*. (Baltimore: Center on Families, Communities, Schools, & Children's Learning, Johns Hopkins University, September 1995): 11.
54. Anne Wescott Dodd and Jean L. Konzal, *Making Our High Schools Better: How Parents and Teachers Can Work Together* (New York: St. Martin's Press, 1999), pp. 187–191.
55. See www.pureparents.org.
56. Karla Scoon Reid, "Black, Hispanic Parents Urged to Support Education," *Education Week* 21, no. 1 (September 5, 2001): 3. More information is available at www.SchoolSuccessInfo.org, or telephone (800) 281–1313.

57. Derrick Smith, "Staying Involved During the High School Years," *Community Update* 91 (September 2001), 3.
58. Jay Mathews, "Blacks Battle Achievement Gap: Parents Unite to Make Sure Children Aren't Shortchanged," *The Washington Post* (December 31, 2000).
59. Joy Dryfoos, "The Mind-Body Building Equation," *Educational Leadership* 57, no. 6 (March 2000): 14–17.
60. Sue Maguire, "A Community School," *Educational Leadership* 57, no. 6 (March 2000): 18–21.
61. Partnership for Family Involvement in Education, "Examples of Families and Schools Working Together."
62. Ibid.

CHAPTER 10

1. The data presented here are from the March 2001 supplement to the Current Population Survey (CPS); for the source of official poverty estimates, see www.census.gov/ftp/pub/hhes/poverty/poverty00/pov00hi.html.
2. United States Pregnancy Rates for Teens, 15–1. National Center for Chronic Disease Prevention and Health Promotion; pregnancy rates from www.teenpregnancy.org/fedprate.htm.
3. S. Jody Heymann and Allison Earle, "Low-Income Parents: How Do Working Conditions Affect Their Opportunity to Help School-Age Children at Risk?" *American Educational Research Journal* 37, no. 4 (2000): 842.
4. The survey was conducted by the U.S. Centers for Disease Control and Prevention and reported in Harriet Tyson, "A Load off the Teachers' Backs," *Phi Delta Kappan*, 80, no. 5 (January 1999): K2.
5. U. S. Department of Education, National Center for Education Statistics, *Digest of Education Statistics*, 2000, Table 106; online at nces.ed.gov/index.html.
6. Reported by the U.S. Department of Education in November 2001. Cited in: "Maine Has Nation's Highest High School Graduation Rate," *Portland Press Herald* (November 16, 2001): 16A.
7. M. Mauer and T. Huling, *Young Black Americans and the Criminal Justice System: Five Years Later* (Washington, DC: The Sentencing Project, 1995) online at www.csdp.org/factbook/racepris.htm+african+american+males+ percent2B+prison&hl=en.
8. Craig Haney and Philip Zimbardo, "The Past and Future of U.S. Prison Policy: Twenty-five Years after the Stanford Prison Experiment," *American Psychologist* 53, no. 7 (July 1998): 716; online at: csdp.org/factbook/racepris.htm+african+american+males+ percent2B+prison&hl=en.
9. National Center for Education Statistics; online at nces.ed.gov/fastfacts/display.asp?id=16.
10. Alvin Toffler, *The Third Wave* (New York: Morrow, 1980), p. 119.
11. See Jane Roland Martin, *The Schoolhome* (Cambridge, MA: Harvard University Press, 1992).
12. David Tyack and Elisabeth Hansot, *Managers of Virtue* (New York: Basic Books, 1982), p. 257.
13. Ibid., p. 261.

14. American poet John Godfrey Saxe (1816–1887) based the poem on an Indian fable. From *Poems for Modern Youth*, Adolph Gillis and William Rose Benet, eds. (New York: Houghton Mifflin, 1938), pp. 190–191.
15. National Center for Education Statistics, *The Condition of Education 2001* (Washington, DC: U.S. Department of Education, 2001), p. 115; online at nces.ed.gov/pubsearch—key in publication number 2001072.
16. Tyack and Hansot, *Managers of Virtue*, p. 254.
17. Sharon L. Kagan and Lynda G. Hallmark, "Early Care and Education Policies in Sweden: Implications For the United States," *Phi Delta Kappan* 83, no. 3 (November, 2001): 241.
18. Ibid., p. 245.
19. Heymann and Earle, "Low Income Parents," p. 842.
20. Ibid., p. 844.
21. Anne C. Lewis, "Time To Talk of Early Childhood," *Phi Delta Kappan* 83, no. 2 (October 2001): 104.
22. Janet Atkin and John Bastiani (with Jackie Goode), *Listening to Parents: An Approach to the Improvement of Home/School Relations* (London: Croom Helm, 1988), p. 17.
23. Tyack and Hansot, *Managers of Virtue*, p. 13.
24. Ibid., p. 260–261.
25. Arthur G. Powell, Eleanor Farrar, and David K. Cohen, *The Shopping Mall High School: Winners and Losers in the Education Marketplace* (Boston: Houghton Mifflin, 1985).
26. Tyack and Hansot, *Managers of Virtue*, p. 257.
27. Ibid., p. 259.
28. Atkin and Bastiani, *Listening to Parents*, pp. 16–17.
29. Toffler, *The Third Wave*, p. xx.
30. Tyack and Hansot, *Managers of Virtue*, p. 260.
31. Carol Merz and Gail Furman, *Community and Schools: Promise and Paradox* (New York: Teachers College Press, 1997), p. 98.
32. Ibid., 98.
33. Jane Roland Martin, "Women, Schools, and Cultural Wealth," in *Women's Philosophies of Education: Thinking Through Our Mothers*, ed. Connie Titone and Karen E. Maloney (Upper Saddle River, NJ: Merrill, 1999), p. 161.
34. John Dewey, *The School and Society* (Chicago: University of Chicago Press 1900/1990), p. 7.

ABOUT THE AUTHORS

THE COLLABORATION

Unknown to each other, Anne and Jean had both chosen to do research about parents and high schools for their doctoral dissertations at different universities. Both studies involved interviewing parents and educators in high schools involved in school reform. A mutual friend put them in touch with each other, and thus the collaboration began. They combined the knowledge gained from their individual studies and did further research to write their first book, *Making Our High Schools Better: How Parents and Educators Can Work Together*, which was published by St. Martin's Press in 1999. Because parents are so central to the success of schools and the achievement of students, they continued their partnership to learn more. Although they individually worked to collect the data for some chapters of this book, the book itself is the result of the many hours they have spent discussing and debating—in person, on the phone, and online—what they found out and what it might mean for public education. Neither Anne nor Jean could have written this book alone, and both believe that working together produced something that is better and more useful than anything they might have done by themselves.

ANNE WESCOTT DODD

I tell my college students and student teachers that I am a dinosaur because I have been involved in education since I graduated from college with a B.A. in History and Government more than four decades ago. After teaching high school English in Maine for one year, I went to Southern California. There I spent six years teaching English, social studies, and conversational French in two junior high schools—both of which included students from very diverse backgrounds—and earned an M.A. in English. Returning to Maine in the late 1960s, I worked as an elementary supervisor and English teacher in a small rural school district and then moved to the more prosperous and populated southern part of the state. I continued to teach English in two high schools and served as acting principal/assistant principal at a high school and principal of a middle school. Since 1984 I have taught education courses and worked with student teachers at Bates College, a liberal arts college in Lewiston.

Throughout the years I have been actively involved in several state and national professional organizations, conducted workshops for teachers in many school districts, and presented my work at state and national conferences. My publications include nine books (six of them in the field of education—including the previously mentioned book coauthored with Jean) and more than 200 articles on educational topics in a variety of general interest publications, such as *The Christian Science Monitor, The Maine Sunday Telegram,* and *The Baltimore Evening Sun,* and in many professional journals, such as *Phi Delta Kappan, Educational Leadership,* and *Education Week. A Parent's Guide to Innovative Education* (Noble Press, 1992) was named one of the Ten Best Books for Parents by *Child* magazine in 1992.

When I began work on my doctorate at the University of Maine in 1990, I already knew I wanted to focus my research on parents. I had seen from my own experience how important parents were to the success and smooth operation of schools, but I also wondered why the educational pendulum kept swinging—from progressive practices to "back-to-basics" approaches. As a middle-school principal, for example, I encountered parents who refused to let their children participate in a five-day outdoor education program at a nearby camp. Instead, each year a few children were forced to sit in a self-contained study hall for a week while all of the other sixth graders were learning about nature firsthand, developing group skills, and building bonds with each other as they camped out and cooked their own food. These parents told me they didn't want their children to "have a vacation on school time." Yet other parents thought this week was the best part of the sixth-grade curriculum. Why did parents think so differently about what counted as "good" educational practices?

When I began looking for answers in the research, I was surprised to find that, although there were many studies on parents of young children, very few had been conducted with parents of secondary students. My dissertation, "Parents as Partners in Learning: Their Beliefs about Effective Practices for Teaching and Learning High School English" (1994), an attempt to address this gap in the research, was named the 1996 Outstanding Dissertation by the Families as Educators Special Interest Group of the American Education Research Association.

As the mother of two adult daughters, one of whom is just beginning her career as a special-education teacher in Massachusetts and the other who has been an active parent volunteer and substitute teacher in Florida, I also have six grandchildren. For many years I have been the "family consultant" about school issues—and there have been many. Because I have intimate knowledge of schools and school systems, I know how the bureaucracy works and what parents need to do to be heard. Thus, I have been able to advise my daughters when they didn't know where to begin. Sadly, most parents have neither the resources nor the knowledge to help them negotiate this unfamiliar terrain when there is a problem with their children. But, too often, even when my daughters benefited from being able to talk through problems with me, educators in their children's schools still refused to listen. I hope that what Jean and I have learned and include in this book will be a wake-up call for everyone.

Although I would suggest different means for achieving it, I wholeheartedly endorse President George W. Bush's goal to "leave no child behind." The schools, however, cannot meet such a complex challenge—made even greater after September 11, 2001—by themselves. Nor should they.

Perhaps educators and some parents can take the lead, but creating a better future for us, our children, and our grandchildren will require the care and commitment of all of us. Just as this book was made possible because Jean and I collaborated, so, too, do I believe that the world can be changed by the collaboration of all of us. Am I too idealistic? too optimistic? Maybe. As I look at the bright, committed, and caring young people I work with as student teachers each year, I believe anything is possible—but only if we give these young teachers the support they need to stay in teaching. I am certain they will make a positive difference in the lives of their students. There are many children, many teachers, many schools across the nation—and there are many adults in every community who could contribute in some small way to the education of our children. What if we all joined hands and worked together?

JEAN L. KONZAL

I come to this work of trying to bridge the gap among teachers, parents, and community members for the benefit of our children from many perspectives: as an educator, a researcher, a parent, and now a grandmother.

After graduating from Queens College's urban teacher preparation program in 1964, I taught in East Harlem in New York City for five years.

Following that I gave birth to my son, Gregory, and after almost two years at home, I moved with my husband, Bill, and our son to New England because we wanted to experience more of the world than just New York. Since then our travels have taken us to New Hampshire, Arizona, Vermont, back to New Hampshire, to Maine, and most recently to Pennsylvania and New Jersey.

As a result of this journey, I have the advantage of multiple perspectives of the American educational landscape. I have taught poor inner-city children in New York and in Tucson. I have taught poor rural children in New Hampshire. I have seen firsthand the disparities between schools in neglected urban centers and schools in wealthy suburbs—sometimes on the same day. I have been a classroom teacher, a reading specialist, an educational consultant on a multidisciplinary team working with children with special needs, a consultant to teachers about mainstreaming, a staff development specialist for a school-university partnership, a state-level school reform consultant, an evaluator of urban school restructuring efforts, and, most recently, a teacher educator.

After working for seven years with Maine's Department of Education, I returned to school and in 1995 earned a doctorate in Policy, Planning, and Evaluation from the University of Pittsburgh. At the time I decided to return to school, I had been working with teachers and principals of high schools in Maine who, in their attempts to change their teaching and learning practices, had run into countless problems. I wanted to gain a deeper understanding of these problems. As I began reading about previous attempts to change schools, I became aware of a relatively unexamined issue, one that resonated with my own experiences as a parent—the role parents play in school change—and decided to examine this issue more fully. In writing my dissertation, "Our Changing Town, Our Changing School: Is Common Ground Possible?" I chose to break with tradition and present my findings in the form of two readers' theater scripts. These scripts, designed to be used to open discussion about educational issues in any community, make my research accessible not only to scholars but also to practitioners, policymakers, and parents. My dissertation was awarded the 1997 Outstanding Dissertation Award by both the Families as Educators Special Interest Group and the Division D/Qualitative Research Special Interest Group of the American Education Research Association.

Parallel to and intertwined with my role as an educator and researcher has been my role as a mother and more recently a grandmother. As a mother, I have struggled to find schools that are congruent with my beliefs about what should go on in "good" schools or to influence the schools that my son attended—and it hasn't been easy. I tried a variety of approaches: one-on-one conversations with my son's teachers and principals, participation on task forces, and volunteering in school. In retrospect, the only way that I actually was able to achieve my goals was to carefully investigate a number of schools and then to move to the community where the schools most closely matched my ideas of a "good" school. It was this strategy that led us to move to Westminster West, Vermont. My son spent his second and third grades in a one-room country school in a town peopled not only by longtime Vermonters but also by a growing number of "flatlanders"—people like me from more urban areas who fled to rural Vermont in search of the good life. This school became my model for a "good" elementary school; the teacher there, Claire Oglesby, my model for a master teacher. My attempts to influence the public schools—so that they looked more like Claire's school—my son attended after this school were not as successful. I found it was not easy for a parent—even one like me with many years as an educator—to influence school teaching and learning practices.

Upon receiving my doctorate, I decided to focus my energies on preparing teachers. As a mother and now a grandmother of three wonderful grandchildren, I know that the most important contribution I can make is to help to prepare teachers with the knowledge, skills, and attitudes to teach all "our" children. So for the past five years I have been a teacher educator in the Department of Elementary and Early Childhood at The College of New Jersey. One of the projects I ask my freshman students to undertake is an interview of a practicing teacher. When asked what is the hardest part of their job, they overwhelmingly tell of problems with parents. This, along with the complaints my student teachers hear from many of their cooperating teachers, confirms what we have long suspected—the parent/teacher relationship is a troubling one for many teachers. The challenge for me as a teacher educator is to find ways to help new and veteran teachers to gain skill in reaching out to parents and to help schools I work in to be inviting to parents.

In the aftermath of September 11, the future of the children of the world, my grandchildren, your children and grandchildren—their education, their security—must take priority. We must each find a way to make a contribution to assure our children's futures. Preparing caring, imaginative, and intellectually curious teachers who I would be happy to have teach my grandchildren is my small contribution.

INDEX

academic program, 227
accountability, for educational development, 152
achievement, student: black-white gap in, 280; and parent involvement, 7
Adams, Kimberly Sue, 141
"Addressing the State of Parent-Teacher Relations" (Murfitt), 143, 149
Adler, Kurt, 99
administrators, 124, 153, 230. *See also* principals
Advertising Council, 280
advisory activities, 116, 173, 175, 213–15, 219, 229
advocacy, 64, 175, 279
aid, for education, 10, 108, 277, 294, 300–302
Alaska, 131
America, Richard, 280
Amherst, Massachusetts, 261
Ammons, Joe, 56–62, 67, 69
Anderson, Don, 271
anger, and trust, 173
Annie E. Casey Foundation, 119
anonymity, 322n.1. *See also* confidentiality
Antarctica, 272
Anwatin Middle School, 269
"[Ap]parent Involvement" (Fine), 121
Arkansas, 42, 58
Arsenault, Kelly, 233
Arts Integration Day, 162–63
ASPIRA Association, Inc. (national), 275
assembly, school community, 188
assessment: grading, 75–77 79–81, 85–87; report cards on parents, 163–64; of schools, 235; for special education, 48, 49, 52, 61–62; standards for parents, 163–64; tests for, 8–9
Atkin, Janet, 114, 168, 296
Ayers, William, 112–13

Balicki, Scott, 115
Baltimore, Maryland, 265
Barth, Roland, 174
Bastiani, John, 114, 168, 296
Bean, Susan F., 236
beliefs, 22, 71. *See also* common ground/purpose; mission; value system; vision
Bellah, Robert, 105
Bensman, David, 161, 170
Berkeley (California) High School, 124, 147
Biggs, Susan Connell, 261–62
Blackburn, John, 90
Black History Month, 204, 206, 218
Black Teachers on Teaching (Foster), 161
"Blind Men and the Elephant, The" (Saxe), 290–92
Bograd, Harriet, 204, 224, 225
Boston University, 269
Boyer Center, 231–32, 242, 251, 253, 263, 271
Boyer, Ernest L. Sr., 231
Boys Club of America, 282
Bozowski, Jan, 191
Brown, Marilyn, 28, 32
Brunswick, Maine, 236
Bucks County Courier Times, 159
Buffalo Grove, Illinois, 131
Buhrer Elementary School, Cleveland, Ohio, 240
bureaucracy, 23–24, 105, 112–13, 249, 298, 303–304

Bush, George W., 335

calendar, school, 187, 220–21, 239
Calhoun, Emily, 128
California, 147
Camden Hills (Maine) Regional High School, 131
Campbell, Robin, 173
Camp Fire USA, 275
Carey, Thomas P., 102
Center on Families, 269
change, 3–4; in beliefs, 22; in democratic society, 17, 307–308; in education environment, 7, 101, 153, 271, 286–87; pitfalls of, 298, 302–303; process of, 9, 40, 70, 91–96, 111, 134–37, 157–59
Chanlatte, Ana, 212, 214, 219
chaos theory, 105–106, 111
Chicago School District, 163–65, 248, 275
Children's Aid Society, 282
Cho, Erica, 191, 192
Christian, Karen M., 266
churches, and schools, 276, 279
Ciotti, Holly, 260
Citizens for Excellence in Education, 84, 86
City College School of Education, New York City, 247
Class Dismissed (Maran), 147–48
Clemens-Brower, Teresa Jo, 238, 255
Coalition of Essential Schools (national), 73, 171, 183, 211
Coffin, Stan, 64
Cole, Jim, 226
college, preparation for, 85, 250
College of New Jersey, The, 183
Collins, Timothy, 130
Colquist, Lori, 194
Comer, James, xx, 252
common ground/purpose, xv, xvii, 128, 229, 241–42, 279, 283, 297–98, 307. *See also* beliefs
communication, 54, 60–61, 94, 166, 305; with community, 240; facilitating, 8–9, 103, 166, 270, 305–306; home-school, 31–32, 114, 157, 232; in paradigms, 31–32, 38–39, 124. *See also* connectedness

communication, via: clubs, 275; conferences, 35, 181, 197–200, 233, 239–40, 245, 266; councils: 116, 182, 187, 192, 223, 270; email, 238; home visits, 181, 197, 198–99, 245–46, 270; journals, 245; letters, 244; network, 278; newsletter, 188, 237, 258; online, 203–206, 224; open houses, 35; partnerships, 140, 259; phone, 214, 238, 267, 272; projects, 194, 197, 199, 245, 274, 303; public forums, 46, 56–57, 81–83, 90–91; seminars, 276–77; surveys, 116, 199, 235, 244–45; teams, 74, 265, 267, 277–78; workshops, 117, 191–92, 247, 258, 276
community, 5, 136, 181, 240, 276; building of, 201–202, 222; connectedness in, 104–10, 118–25, 135–36; diversity of, 203–208, 218, 222, 224–25, 227–29; education in, xvi–xvii; in paradigm, 38–41. *See also* culture; educators; family
community home, 104, 288, 289
confidentiality, xxi, 44, 58, 59, 61, 245, 322n.1
conflict: as opportunity, 296–97; resolution of, 173–74
connectedness, 104–9, 118–26, 135–36. *See also* communication; community; involvement; unity
Consolidated School, Kennebunkport, Maine, 257
Cooper, Stan, 55
Crandall, Patricia, 56–58, 59–60, 62–64, 66–67
Creamer, Clyde "Buck," 207
Cremin, Lawrence, 27
criminal justice system, 43–45, 61, 150, 286
Cross Country Elementary School, Baltimore, 265, 269
Crossroads School, Manhattan, xxii, 173, 176, 203–31, 273; advisory program of, 213–15, 219, 229; director of, 155, 205–207, 216–18, 222, 227, 230; discipline at, 212–13; diverse community of, 207–208, 222, 225, 227–29; leadership of, 210–11,

218–21, 223; parents at, 160, 225–27; racism issue at, 203–6; as school of choice, 207–10, 215; trust building at, 167

culture, 14; African American, 124, 147–48, 204, 206, 218, 227, 279–80, 286; Asian, 280; of community, 127–28; differences in, 161, 163, 170–71, 201, 205, 206; Latino, 224, 279–80; minority, 147, 279–80, 290, 297; Native American, 240–41, 280; in paradigm transition, 127–28; school, 89; Swedish, 294. *See also* community; diversity

curriculum, 18, 27–30, 73–79, 103, 206, 211, 216

Curtis, Carissa, 79, 81–82, 90

Cutler, William, 23, 36, 108

Dahlberg, Thea, 155, 194, 197, 200

Dallas (Texas) Area Interfaith Group, 279

"Damage Control" (Maran), 148

D'Amato, Grace, 55

Danebo Elementary School, Eugene, Oregon, 231, 242, 251

Davis, Diane, 43–47, 49–52, 56–58, 64–65

DeBlois, Larry, 175, 239

decision making, 6, 37–38, 150, 182, 192, 225–27, 265, 279, 299–300

democracy, 6, 12, 287, 307–308

demographics, 13–14, 161, 182, 208, 222, 285–86. *See also* diversity

Deschambeault, Debbie, 233

Desmond, Carol, 183, 185, 186, 189, 193

Dewey, John, 6, 124, 308

Dezan, Beverly, 191

director. *See* principal

disabilities, learning, 8–9, 15, 53

discipline, 212–13, 270

Discovery Elementary School, Buffalo, Minnesota, 231, 250, 263

diversity, xvi, 121, 124, 297; of community, 203–208, 218, 222, 224–25, 227–29; of culture, 161, 163, 170–71, 201; of teachers, 208, 282. *See also* common ground

Dodd, Anne Wescott, xvi, xix, 333–35

Dorson, Mitch, 259

due process, xxi, 43–46, 61, 67, 150, 186. *See also* criminal justice system; police

Dundas, Brendan, 244

Earle, Allison, 295

East Harlem, New York, 276

Eastman Kodak Company, 280

education, xvi-xvii, 17, 152, 235, 238, 285–87; in contrast to schooling, 4–5, 136; knowledge base for, 243, 290, 297–98; public, 290, 300–301, 305, 306; resources for, 10, 208, 277, 294, 300–301, 302. *See also* learning; special education

Education Week, 163

educators, 8–9, 37–38, 166–67, 284, 292–93; trust issues for, 140–43, 149–54; 157–158. *See also* administrators; principals; teachers

Edwards, Patricia A., 243

Ehrich, Terry, 253

Eli Whitney School, Enfield, Connecticut, 248

Elk Grove (California) Unified School District, 282

English, 114, 117, 250, 260

Epstein, Joyce, xx, 107, 168

Ernest, Paul, 22

Errol Hassell Elementary School, Aloha, Oregon, 238

Escambia County, Florida, 277

"Essential Collaborators" (Coalition of Essential Schools), 171

Etzioni, Amatai, 105

Evans, Robert, 94

events, recreational, 173, 186, 188, 220–22, 245, 268; school schedule of, 187, 220–21, 239

"Evoking the Spiritual in Public Education" (Palmer), 136

Fairfield Court Elementary, 270

Families and Schools Together (FAST), 250

family, 13–14, 27–30, 113–17, 243, 249, 277. *See also* parents

Family Service, 250

Fergoson-Florissant (Missouri) School District, 270
Fienberg-Fisher Elementary School, Miami Beach, Florida, 278
Fifth Discipline, The (Senge), 120
Fine, Michelle, 40, 121, 124
First and Central Presbyterian Church, Wilmington, Delaware, 276
Fisher, Annette, 57, 64–65
Flaxman, Erwin, 119, 129
"Following Their Kids to School" (Ginn), 19
Foothills High School, Tucson, Arizona, 259
Foster, Michelle, 161
Fowler, Nancy, 244
Framework for Understanding Poverty, A (Payne), 242
Freeport (Maine) Middle School, 267–68
Friedman, Robert, 7
Fullan, Michael, 95
Furman, Gail, 306

Gardner, J. W., 127
Gehm, Vickie, 8
Geraldine Palmer Elementary School, Pharr, Texas, 130–31
Gill, Janet, 191
Ginn, Linda, 19
Girls Club of America, 282
Glenallan (Maryland) Elementary School, 280
Glendale (California) High School, 260
Goethe School, Logan Square, Illinois, 269
Goldberg, Nancy, 169, 257
Good, Howard, 23
Gordon, Ellen, 188, 191
Gordon, Ira J., 106
government, 107, 112; as funding source, 10, 108, 294; school, 92
Green, Thomas, 22
Greenbrook (New Jersey) Elementary School (GES), xxii, 143–44, 155, 176, 181–202; community of, 171, 176, 181–83, 201–202; conferences of, 197–200; parents in classroom, 193–97; parents of, 186–92; physical space of, 184; principal of, 183, 185–86, 189–90, 193, 200–201; site council of, 182, 187, 192
Gregg, Kathy, 244

Hallmark, Lynda, 293
Hansen, Angela, 280
Hansot, Elisabeth, 290, 292, 297, 298, 301, 305
Hargreaves, Andy, 17, 116
health issues, 278
Heath, Shirley Brice, 28
Henry, Mary, 105, 168, 184, 225
Herman, Minnesota, 271
Heymann, S. Jody, 295
history, for parents, 259
Holliday, Patricia, 155, 176, 183, 185–86, 189–90, 200–201
Holmes, John G., 142
home: education in, xvi-xviii, 13; literacy in, 243. *See also* children; family; parents
homework, 244, 248
Hooper, Sharon, 272
Horton, Margie L., 122
Horwath, Chris, 256
Houston, Paul, 24, 105, 107

IEP (Individualized Educational Plan), 47, 48, 49, 52, 63
Illinois Family Education Center, 249, 250, 259
imagination, in paradigm transition, 132–37
information, 286–87. *See also* knowledge
Institute for Responsive Education, 269
intellectual development, 102–103
interaction, in paradigm transition, 104–109. *See also* communication; community
Invisible Privilege (Rothenberg), 205
involvement, parent, 32–36, 191, 193, 256, 268, 278–80; handbook on, 106; as partners in teaching, 27, 113–14, 117, 236, 267–70, 272–73; on portfolio projects, 233–34; in reading, 188–89, 195–96, 254–55; in schoolwide activities, 186–88; value

of, 7, 14, 107, 119, 253, 257. *See also* communication; parents
Iowa Department of Education, 106

Jackson, Beth, 63, 65, 69
Jackson, Tennessee, 276
Jacobsen, Wayne, 241
James Mosher Elementary School, Baltimore, 265, 269
Jervis, Kathe, 207
Jordan, Carolyn, 24, 203
Joyce, Bruce, 128
Juniewicz, Kit, 274
Just, Diana, 239

Kagan, Sharon, 293
Kamon, Dean, 274
Kerber, Julie Macosko, 254
Kipps Elementary School, Blacksburg, Virginia, 231, 242, 271
knowledge, 243, 290. *See also* information
Konzal, Jean, 2, 19, 149, 153, 172, 326nn. 19, 335–37, xix, xvi

labeling, *vs.* problem solving approach, 295
Lally, Miriam, 185, 192, 194, 202
language, 34–35, 114, 117, 250, 260
Las Cruces, New Mexico, 278
leaders, 160, 218–19, 220–21
leadership, 155–60, 165–69, 174–76, 217–21, 273
Leadership and the New Science (Wheatley), 99
Lear, Rick, 171
learning, 118, 127–28, 177, 189–92, 275, 292. *See also* education; schooling
learning disabilities, 8–9, 15, 53
Levy, Harold O., 235
Lewis, Anne C., 295
Lieberman, Linda, 150
Lincoln, Illinois, 270
Lincoln Middle School, Portland, Maine, 233
linear thinking, 303
literacy, 195–96, 243, 282
"Load off the Teachers' Backs, A" (Tyson), 104

Lowell, Massachusetts, 130
Ludden, Julie Ann, 238

Madison, Wisconsin, 250, 280
Magliaro, Elaine M., 264
Maine, 131, 233
Making Our High Schools Better (Dodd and Konzal), xix, 333
Malcolm Bell Middle School, Marblehead, Massachusetts, 264
Malu, Kathleen, 228
Manchester, New Hampshire, 274–75
Manhattan, 203, 207, 208. *See also* New York City
Manning, Sue, 244
Maran, Meredith, 147–48
Maranacook Community School, 175
Maree G. Farring Elementary School, Baltimore, 265
Margolis, Lois, 191
Marston, Janice C., 162
Martin, Jane Roland, 102–104, 288, 293, 307
Martin Luther King celebration, 206, 225
Maryland School Performance Assessment Program, 235, 280–81
math, 192, 248, 262
McCaleb, Sudia Paloma, 246
McDonald, Lynn, 250
McGeeney, Patrick, 30
mentoring, 247, 270. *See also* advisory
Merz, Carol, 306
Michel, Kathy Teel, 166
Miller, B. A., 130
Minnich, Elizabeth Kamarck, 132
minorities, 147, 279–80, 290, 297
Mishra, A. K., 140
mission statement, 73–74, 102, 223. *See also* beliefs
model: *see* paradigm; satellitic; synergistic
Modesto (California) School District, 241
Molly Stark School, Bennington, Vermont, 281–82
Monohan, Pat, 254
Montello School, Lewiston, Maine, 258, 262, 263

Moore, Marjorie, 203, 226
Moyer, Betsy, 33
Mozart School, Logan Square, Illinois, 250
Mt. Sinai Health Care Center, 276
Mt. Vernon (Virginia) High School, 240
Murfitt, Kathy, 143, 149

Nadeau, Deanna, 258, 262, 263
Nation at Risk, A, 5
National Association for the Advancement of Colored People (NAACP), 280
National Center for Educational Statistics, 127
National Parent Teachers Association (PTA), 107
National Research Council, 295
National School Age Child Care Alliance, 185
National Semiconductor, 274, 275
New Brunswick (New Jersey) High School, 278
New England, 42
New Haven, Connecticut, 152
New Jersey, 182–83, 278
New Meaning of Educational Change, The (Fullan), 95
New York City, 207, 208, 209, 216, 230, 235. *See also* Crossroads School; Manhattan
Newark, New Jersey, 122
Nichols-Solomon, Nicole, 235, 242
Northeast Middle School, 269
Northwestern University, Chicago, 276

Oglesby, Claire, 172, 337
"Open Letter from a Parent, A" (Zukas), 20
opportunities, for interaction, 26, 35, 81, 116–17, 225, 227; conflict as, 296–97; and trust-building, 161–64, 171–73
O'Shea, Cindy, 167
"Our Changing Town, Our Changing School" (Konzal), 19, 149, 336

Paige, Rod, xv, xvii
Pajares, Frank, 22
Palmer, Parker, 136

paradigm, new, xx, 191, 304–307. *See also* synergistic model
paradigm, traditional/old, xix, xxi, 18–41; bureaucracy in, 23–24; communication in, 31–32; community in, 37–41; curriculum in, 27–30; my child *vs.* all children in, 22–23; parent involvement in, 32–36; philosophical differences in, 21–22; satellitic model of, 24–25–27
paradigm, transitional, 40, 99–137; connectedness in, 104–109, 118–26; culture in, 113–17, 127–28; development process of, 110–13; imagination in, 132–37; schooling and education in, 100–104, 113–17, 129–32; tradition in, 99–100
Parent and Community Training Academy (PCTA), Chicago Public Schools, 248–49
Parental Involvement and the Political Principle (Sarason), 139
parents: concerns of, 8–9, 95–96, 166; in decision making, 37, 150, 182, 225–27, 265, 279; income of, 286, 294; of kindergarten children, 189–91; as leaders, 160, 220–21; as learners, 53, 189–92; organization of, 220–22, 224, 236; orientation for, 239; in paradigms, 32–36, 191; research on, 333–34, 336; research with teachers, 269–70; roles of, 158–60, 163–64, 218, 247, 284; support to, 293–95; trust issues of, 140–43, 153–54, 215; working mother, 286. *See also* family; involvement
"Parents as Partners in Learning" (Dodd), 334
Parents of Children of African Descent, 124, 147–48
Parents United for Responsible Education (PURE), 163–65, 279
parent-teacher organization, 107, 188, 240
parent/teacher relationship, 337
Partnership for Family Involvement in Education (U.S. Department of Education), 107

Patrick O'Hearn School, 270
Payne, Ruby, 242
People for the American Way Foundation, 280
Perkins, Dennis, 79–81, 84–86, 87–89, 91–95
personal/social development, 102–103
Personal Vision of a Good School, A (Barth), 174
Perstein, Linda, 143
Pharr, Texas, 130–31
philosophy: of education, 297–98; school, 210–11; teaching, 266; in traditional paradigm, 21–22
photography, 247
Planetary Studies Foundation, 272
planning: non-linear, 110–11; school, 70–74
Pleasant, Heather M., 243
poetry, 264
Pohlig, Colleen, 173
police, and schools, 42–43, 50, 56, 57, 58–59. *See also* due process; safety
politics, 10, 201, 290, 298–99. *See also* bureaucracy; government; referendum
Pontiac, Michigan, 131
poverty, 286
power, 40, 107, 124, 150–51, 298–300, 302
principals: administrative issues of, 230; on parent partnership, 267; role of, 153, 155–57, 216–18. *See also* Ammons; Crandall; Perkins; Holliday; Wiener
privacy, 245. *See also* anonymity; pseudonyms
problem-solving approach, 295–96, 299
Professional Development School Network, 183
professionalism, 20–21, 34–35, 145–46, 149–51, 153, 168, 170
"Promise of Urban Schooling, The" (Flaxman), 129
pseudonyms, xxi, 322n. 1. *See also* confidentiality
psychiatrist, evaluation by, 62
psychology, 95
PTA, 107, 188, 240

Public Education Fund Network, 240
public relations, 85
purpose, common, 128, 229, 242, 279, 283, 297–98, 307. *See also* beliefs

questionnaires: *see* surveys
questions, for change, 134–36

race, education-related statistics by, 286. *See also* culture
racism, 203–206
Rampel, John K., 140–43
reading, 188–89, 195–96, 254–55, 258–59
referendum, public, 10, 75, 83, 85–86, 93. *See also* politics
reform, school, 7, 153, 271. *See also* change
Reich, Jesse, 322n.1
Reisman, Mona, 197
research: action, 269–70, 303; on parents, 333–34, 336
respect, xxii, 139–40, 142. *See also* trust
responsibilities: balance of, 229; compact of, 270
Retirees Enhancing Science Education through Experiments and Demonstrations (RE-SEED), 272–73
Rokeach, Milton, 22
Roosevelt High School, Dallas, Texas, 250, 279
Rothenberg, Paula, 205
Rothstein, Shari, 181, 185, 193, 196
Roundtable, 78

Sacramento City (California) Unified School District, 245
Sadler, Brenda, 47
safety, school, 42, 58–59, 102, 143–44, 327n.5
Samuel Gompers Fine Arts Option School, Chicago, 270
San Franscisco, 246
Sarason, Seymour, 139, 265
satellitic model, 24–25–27, 38–39. *See also* paradigm, old
Saxe, Geoffrey, 290–92
schedule, 157, 187, 220–21, 239
School and Society, The (Dewey), 6

school board, 89
School Development Program (national), 252
schoolhome, 103–104, 288
schooling: in contrast to education, 4–5, 136; in paradigm transition, 100–104
School Leadership Team, 220
schools: charter, 130, 304; of choice, 207–10, 215, 263; community, 127–32, 129–32, 271, 281; curriculum of, 18, 27–30, 73–79, 103, 206, 211, 216; education in, xvi–xvii, 4; environment of, 123, 131–32, 184, 251–52, 268, 305; governance structure of, 92; healthy, 278; high, 256, 260–61; home, xvi–xvii, 13; kindergarten, 169–70, 181–82, 189–91, 266; middle, 267–68; off-hours use of, 185, 276, 282; orientation to, 106, 181, 189–91, 213, 239; in paradigm transition, 100–104, 113–17; philosophy of, 210–11; progressive, 210; public, 4–5, 12; resources for, 10, 108, 277, 294, 300–302, 305; special-purpose, 50
"Schools Spelling Out Need for Parental Civility" (Pohlig), 173
science, 192, 236, 272–73
Scott-George, Katrina, 147, 148
Sears, Nancy, 191
Sendak, Judith, 82, 86–87
Senge, Peter, 120, 155
September 11, 2001: issues after, xv, xvii, 138–39, 287, 307, 327n. 5. *See also* safety
Sergiovanni, Thomas, 217
service learning, 118
settings, interaction, 305. *See also* environment
sharing, lesson in, 159
Sherwood Heights Elementary School, Auburn, Maine, 162
Shiloh Baptist Church, Washington, D.C., 276
Sipiera, Paul, 272
Sisko, Ann, 194
Sizer, Nancy Foust, 103
Sizer, Theodore, 103
Smith, Derrick, 13

Snohomish County, Washington, 279
social-emotional development, 211
socialism, classroom, 159
social justice, curriculum based on, 211, 216
social movement, for educational change, 17
social/personal development, 102–103
South Lakes High School, Fairfax County, Maryland, 280
South Portland (Maine) High School, 274
South Side Elementary School, Johnson City, Tennessee, 231, 251
South Wellington schools, 42–69; administrator of, 56–62, 67, 69; due process at, xxi, 43–46, 61, 67; public forums on, 46, 56–57; results at, 62–67; safety at, 42, 58–59; special education at, 43, 45–47, 49–52, 53–56, 61–65, 68–69
special education, 16, 112, 277–78, 286. *See also* learning disability; South Wellington schools
Spielman, Jane, 247
Spinner, Connie, 240–41
Springfield Regional High School (SRHS), xxi, 70–96; change process at, 70, 91–96; curriculum at, 73–78, 79; first two years at, 79–84, 87–91; planning for, 70–74; principal of, 79–81, 84–86, 87–89, 91–95; referendum on, 70, 83, 85–86, 93
St. Agatha Academy, Winchester, Kentucky, 254
statistics, education-related, 285–86. *See also* demographics; diversity
storytelling, 194
students/children: achievement of, 7; early education of, 293–95; honors, 87–88; minority, 280; my child *vs.* all of our, 68, 154, 213, 289, 304, 308; needs of, 16, 33, 43, 63, 133, 307–308; in paradigm, 18–20; role in trust and respect, 157; as volunteers, 118. *See also* family; schools
Style, Emily, 113, 211
"Suspicious Minds" (Perstein), 143
Sweden, 294

synergistic model, xx, xxi, 125–37; culture in, 127–28; imagination in, 132–37; interdependence in, 125–26; schools in, 129–32. *See also* paradigm, new and transitional

Taffel, Ron, 213
Tazlo, Ervin, 106
teachers: diversity of, 208, 282; as leaders, 218–19; parents as, 255; parents in partnership with, 27, 113–14, 236, 269–70; professionalism of, 20–21, 145–46; recognition of, 65–66; trust issues of, 93–94, 140–43, 149–53, 157–58
teaching: methods of, 11–12, 146, 151; philosophy of, 266
technology, 192, 234, 287
Texas Alliance Schools, 131
theory, chaos, 105–106, 111
Theriault, Janet Russell, 267
Third Wave, The (Toffler), 99, 307
3 Cs and 3 Rs, 102–103, 289
Toffler, Alvin, 99, 106, 287, 304, 307
Toy, Chris, 267
tradition, 99–100
traditional paradigm. *See* paradigm, traditional
training, for volunteers, 272–73
Transforming Knowledge (Minnich), 132
trust, and respect, xxii, 94, 138–39, 138–77; building of, 155–60, 165–69, 173–77; with common purpose, 283; for cultural differences, 170–71; educators' role in, 140–43, 149–54, 157–58, 215; opportunities for, 161–64, 171–73; parent's role in, 117–19, 140–43, 143–48, 153–54, 158–60, 215; student's role in, 157
tutoring, 276, 277. *See also* involvement
Tyack, David, 290, 292, 297, 298, 301, 305
Tyner Academy, Chattanooga, Tennessee, 256
Tyson, Harriet, 104

unity, xvi, xvii, 307. *See also* connectedness
University of California, Berkeley, 248
University of Minnesota at Morris, 271
U.S. Census, 13, 112, 222
U.S. Department of Education, 7, 107

Valesco, Elena, 143
value system, 27–28. *See also* beliefs
Vanderall, Joelle, 117, 161
Veazie Elementary School, Providence, Rhode Island, 278
Vermont, 337
violence, in schools, 42, 102. *See also* police; safety
vision, 95, 119–21, 217–18. *See also* belief; change
volunteerism, 265. *See also* involvement

Walden, Virginia, 281
Washington-Monroe Elementary School, Lincoln, Illinois, 270
Weaver, Beth, 143
Webber, Helen, 276
Weber, Sally, 249
Westminster West Elementary School, Vermont, 172
Wheatley, Margaret F., 99
White, Moses, 185
Whitehead, Tony, 280
Wiener, Ann, 155, 205–207, 216–18, 222, 227, 230
Wilhite, Jack, 10
Williams, Bonita, 197, 199
Woestehoff, Julie, 163
Wolaver, Beth, 145
Woods, Lori, 168, 191, 194, 195, 197, 198
Woolsey, Patricia, 149
writing, 187, 189–91, 246, 260, 262

Xu, Jianzhoong, 215

Young, Michael, 30

Zukas, Tim, 20, 156
Zweig, Eileen, 188

GPSR Compliance

The European Union's (EU) General Product Safety Regulation (GPSR) is a set of rules that requires consumer products to be safe and our obligations to ensure this.

If you have any concerns about our products, you can contact us on

ProductSafety@springernature.com

In case Publisher is established outside the EU, the EU authorized representative is:

Springer Nature Customer Service Center GmbH
Europaplatz 3
69115 Heidelberg, Germany